D1633033

NALA'S WORLD

DEAN NICHOLSON

WITH
GARRY JENKINS

**HODDER &
STOUGHTON**

First published in Great Britain in 2020 by Hodder & Stoughton
An Hachette UK company

1

Copyright © Dean Nicholson and Connected Content Limited 2020

The right of Dean Nicholson and Garry Jenkins to be identified as
the Authors of the Work has been asserted by them in accordance
with the Copyright, Designs and Patents Act 1988.

Illustrations © Kelly Ulrich

All rights reserved. No part of this publication may be reproduced, stored
in a retrieval system, or transmitted, in any form or by any means without the
prior written permission of the publisher, nor be otherwise circulated in any form
of binding or cover other than that in which it is published and without a similar
condition being imposed on the subsequent purchaser.

A CIP catalogue record for this title is available from the British Library

Hardback ISBN 978 1 529 32798 4
Trade Paperback ISBN 978 1 529 32799 1
eBook ISBN 978 1 529 32801 1

Typeset in Plantin Light by Palimpsest Book Production Ltd, Falkirk, Stirlingshire

Printed and bound in Great Britain by Clays Ltd, Elcograf S.p.A.

Hodder & Stoughton policy is to use papers that are natural, renewable
and recyclable products and made from wood grown in sustainable forests.
The logging and manufacturing processes are expected to conform
to the environmental regulations of the country of origin.

Hodder & Stoughton Ltd
Carmelite House
50 Victoria Embankment
London EC4Y 0DZ

www.hodder.co.uk

*Sometimes what you're looking for
comes when you're not looking at all.*
Anonymous

What greater gift than the love of a cat.
Charles Dickens

CONTENTS

Bulgaria

PART ONE

Trebinje, Bosnia to
Athens, Greece

PART ONE
FINDING THE ROAD
Bosnia—Montenegro— Albania—Greece

1
Come Home

There is a wise old saying where I come from in Scotland: *Whit's fur ye'll no go past ye.* Some things in life are destined to happen. What's meant to be, is meant to be. It's fate.

From the beginning I had a feeling that is what brought Nala and me together. It couldn't have been a coincidence that we were at the same remote place at the exact same time. Or that she arrived in my life at such a perfect moment. It was as if she'd been sent to give me the direction and purpose I'd been missing. I can never know, of course, but I like to think I brought Nala what she was searching for, too. The more I think about it, the more convinced I become. For both of us, our friendship was simply meant to be. We were destined to grow up and see the world together.

Three months before we met, in September 2018, I had set off from my hometown of Dunbar on the eastern coast of Scotland to cycle around the globe. I'd not long turned thirty and wanted to shake myself free from the routine of my life, to escape my little corner of the world and achieve something worthwhile. It's fair to say it had not been going according to plan. I'd made it through northern Europe, but my journey had been a series of detours and distractions, false starts and setbacks, most of them self-inflicted. I'd planned on completing the trip with a friend, Ricky, but he'd turned around and gone home already. His departure was probably a good thing, if I'm honest. We were not the best influence on each other.

By the first week of December, as I cycled through southern Bosnia *en route* for Montenegro, Albania and Greece, I began to feel as if I was finally making progress. I was ready to have the experience I'd wanted. Long-term, I dreamed of making it through Asia Minor and along the ancient Silk Road and into Southeast Asia, from there down to Australia, across the Pacific and up through South, Central and North America. I pictured myself cycling through paddy fields in Vietnam and across deserts in California, through mountain passes in the Urals and along beaches in Brazil. The world was my oyster. The journey would take me as long as it took. I had set no timetable for myself. I didn't need one; I had no one to answer to anymore.

On that particular morning, I packed up my tent in a small village near Trebinje as day was breaking, around seven-thirty. Aside from a few barking dogs and a dustcart, the shiny cobbled streets were almost empty. I bumped across the stones, the rattling of my off-white bike shaking me out of my sleep, then set off on the road that led up into the mountains and the border into Montenegro.

Snow and sleet showers were forecast for the next day or so, but the skies were clear and the temperatures mild. I was soon making solid progress. After a frustrating few weeks, it felt good to be back on the road and cycling freely. I'd spent much of the past week in plaster, recovering from the leg injury that I'd picked up jumping off the famous Stari Most bridge in Mostar, a few hours back down the road in Bosnia. It had probably been a foolish thing to do. The locals had advised against it during winter when the river was running deep. But I've been prone to doing foolish things all my life; once a class clown, always a class clown.

As far as I was concerned, my big mistake was to listen to the guide, who persuaded me to use a different technique than the one I used to jump off the cliffs back home in Dunbar. I'd hit the freezing cold water with my legs slightly bent. I knew I'd done something wrong the moment I came out of the river. A doctor told me I'd torn the anterior cruciate ligament – or ACL – in my right knee and would need to remain in a plaster cast for three weeks.

I'd cut it off after only one. I'd been too impatient to hang around for longer and I left Mostar before my next appointment at the hospital. So as the sun rose up ahead of me and I made my way up the long, slow climb into the mountains that morning, my main concern was the same as it had been since I'd returned to the road – not to inflame the injury in any way. I knew that my knee was okay as long as I didn't move it from side to side.

I was focused on pumping my legs rhythmically up and down in the same plane. I soon settled into a nice routine, and it seemed to be going fine. I was feeling confident that I might get fifty or even a hundred miles under my belt.

By mid-morning, I had entered a mountainous region on the southern tip of Bosnia. It felt a long way from civilisation. The last town of any size had been ten miles further back down the road. I'd ridden past a quarry of some kind a few miles later, but it was deserted. I was on my own. The spiralling road wasn't especially steep; it was more of a long, slow, gradual climb, which suited me fine. There were sections where the road fell away, too, giving me much appreciated breaks from cycling. The views were spectacular; I was riding along high ridges and looking up to soaring, snowy peaks. It was exhilarating.

I felt so good, I decided to put on some music. The sound of 'Come Home', a new song by one of my favourite artists, Amy MacDonald, was soon blaring out from the speakers I had strapped to the back of my bike. I must have been in high spirits, because I began singing along to the chorus.

On another day, the lyrics might have been designed to make me feel homesick. And there was a moment when I did think about my mum and dad and my sister back in Scotland, waiting for me to come home one day soon. We were a close family and I missed them, but I was enjoying myself too much to dwell on it.

Home will have to wait a while longer, I thought to myself. Of course, it never occurred to me that something else might be waiting for me, a little closer than home.

I was riding along another gently rising section of road when it happened. At first, I wasn't quite able to make out the faint, slightly high-pitched noise that seemed to be coming from behind me. For a moment I dismissed it as the squeaking of my rear wheel or a loose fitting on the bulky panniers that held most of my clothes and other gear. I'd apply some oil when I had my next break. But then, as I stopped singing and the sound became clearer, I realised what it was. I did a double-take. It couldn't be, could it?

It was meowing.

I turned and, out of the corner of my eye, I saw it. A scrawny, grey-and-white kitten was scampering along the road, desperately trying to keep up with me.

I hit the brakes and pulled up to a stop. I was shocked.

'What the hell are you doing here?' I said.

Further down the road, the hills had been dotted with small goat sheds and farms, but high up here in the mountains, I'd not seen another building for miles. There was barely any traffic around. I couldn't fathom where it had come from, or more to the point, where it was going.

I decided to take a closer look, but by the time I'd parked the bike and climbed off, the cat had skipped off the road, through the metal crash barriers and into some loose boulders. I scrambled down and drew close. It seemed obvious to me that it was a kitten, no more than a few weeks old perhaps. It was a scrappy wee thing. It had a long, slim frame, big, sharply pointing ears, spindly legs and a thick tail. Its coat was thin and weather-beaten and was flecked with hints of rusty red. But it also had the most piercing, huge, green eyes, which were staring at me as if trying to work out who I was.

I approached, half expecting it to be feral and to run off when I got nearer. But it didn't seem worried by me in the slightest. It let me stroke it on the back of the neck, leaning into me and purring lightly, as if it was grateful for the human contact and attention.

This cat lived in a normal home, I thought. Perhaps it had escaped or, more likely, been abandoned here at the roadside. I could feel myself growing angry at the thought of it. I could also sense my defences crumbling.

'Poor wee thing,' I said.

I went back to the bike and opened up one of the panniers. I didn't have much food on board, but I decided to spoon out some pesto that I had brought with me for lunch. I smeared the chunky, red paste on to a rock and let the kitten dive in.

It did so as if it hadn't seen food in a week, absolutely wolfing it down. I'd been posting the highlights of my journey mainly for friends and family on Instagram and decided to film this odd encounter on my phone. I might share it with them all later. The kitten was certainly photogenic and it almost seemed to be playing to the camera as it scuttled around the stones at the roadside.

The truth, unfortunately, was far less pretty. Left up here to its own devices, it would die from cold or starvation. It could get run over by one of the giant lorries that occasionally went past. Or even get taken by one of the birds of prey that I'd seen hovering over the mountain peaks. It was so small and delicate, an eagle or buzzard would easily be able to swoop in and pick it up.

I've had a soft spot for animals since I was a kid back in Scotland and have always been drawn to waifs and strays. At various points I've kept gerbils, chickens, snakes, fish, even stick insects. Once, while still at school, I raised an injured young seagull for seven weeks during the summer holidays. The bird became almost tame and my mum and dad still have a photo of me walking around with it on my shoulder. It flew off eventually, healed and healthy, on the day before I went back to school.

Animals being animals, my efforts to help them didn't always pan out. While working on a farm, I made the mistake of taking home a pair of piglets whose mother had died. I put them in my bedroom under some headlights to keep them warm. What an idiot. They ran riot, burying themselves in my clothes and making a mess everywhere. As for the noise – from the squealing sounds they made, you'd have thought they were being murdered. It was the worst night of my life.

I'd always thought of myself as more of a dog than a cat person. I imagined them to be aggressive creatures, but this one looked vulnerable and innocent; it wouldn't hurt a fly. But, while my heart was telling me to pick the kitten up, my head was saying something more sensible. My trip had already contained enough drama and I'd now got some momentum going. If I was going to make it to Montenegro by tonight, I couldn't let this slow me down.

I got back on to the road and simply pushed my bike along, letting the kitten run alongside me. I don't know why, but I felt certain it would soon get bored, see something else to play with and scamper off. But after five minutes or so, it was plain to see it wasn't going anywhere. More to the point, it didn't have anywhere to go. The rocky, scrub-lined landscape was harsh and, if the weather forecasts were to be believed, might soon be covered in snow. It wouldn't last a day up here, I reckoned. If that.

I sighed. My heart had overruled my head. There was no other option.

I lifted the kitten up and carried it over to the bike. It fitted easily into the palm of my hand and weighed next to nothing. I could feel its ribs poking through. I had a 'tech' pouch at the front, where I kept the drone I used to make videos and take photographs of my journey. I cleared it out and put it in one of the panniers.

I then put a T-shirt in the pouch for a lining and placed the kitten gently inside. Its little face poked out, looking at me uneasily, as if trying to tell me it was not comfortable. But there was nothing more I could do. Where else could I put it? I got going, hoping it would settle, but it was soon clear the kitten had other ideas.

I'd barely gone a few hundred yards when it caught me by complete surprise. Before I could do anything, it jumped out of the pouch then crawled up my arm and scampered onto the back of my neck. It then made itself at home. I felt it wrapped around me, its head nuzzled into the nape of my neck, breathing gently. It wasn't uncomfortable or distracting in any way; it was a nice feeling, truth be told. It was obvious the kitten was comfortable too, so I pressed on. Soon, to my amazement, it was fast asleep.

This gave me a breather. A chance to take stock and decide what to do now. I was immediately torn again. On the one hand, while I'd been enjoying being on my own, it was good to have some company. The kitten was hardly a heavy load. It would be entertaining – of that there was no doubt. But on the other hand, it wasn't part of my plan. I'd had too many moments like this, I scolded myself. I was being distracted again.

As mid-day approached the sun was still climbing into the grey-blue sky. I knew from my GPS that the border would soon be drawing near. I'd have to make some decisions. Big ones.

Deep down, though, I had a suspicion that I'd made the main one already.

Whit's fur ye'll no go past ye.

It was fate.

2

The Stowaway

It took another hour and a half to reach the border, but my new passenger remained clamped to my shoulders throughout, snoozing away, totally oblivious to everything. If only I could have felt so relaxed about life right now.

Weaving my way up the mountain road, my mind was working overtime. I was sure that I'd done the right thing. I couldn't have left a vulnerable little creature in such a dangerous place. But at the same time, I was being nagged by doubts. What was I going to do when I reached the border crossing? And what was my plan beyond that? I hadn't bargained on having a cat as a co-pilot.

I briefly convinced myself that I should declare the kitten to the authorities. I'd be honest and simply explain what had happened. I'd found it at the roadside and was going to take it to a vet. Surely they'd be sympathetic? It wasn't as if I was trying to bring through something sinister. It was only a kitten, for heaven's sake. But when I thought it through, I realised that wasn't going to work. There was a reason that every country had rules on the movement of animals. They could carry diseases across borders and kittens were notorious for picking up illnesses. It might need to be put into quarantine; they might even have to euthanise it. I really didn't want that to happen.

For a while, I thought about getting around this by claiming it was my own cat. But I had no paperwork, no medical certificates to vouch for its health. So that was a non-starter too.

I saw that my only option was somehow to slip the cat into Montenegro unseen. I'd work out my next move after that.

I passed a sign: five kilometres to the border. I stopped at a lay-by on the side of the road. A part of me was still hoping to find a loophole, a workaround. So, as a last throw of the dice, I pulled up a map on my phone; there might be a small mountain road or trail that didn't have border guards. But the map showed there was no other route into Montenegro. And besides, it was a stupid idea. What would happen if I was pulled over by police and had no official record of entering the country?

Get real, Dean, I told myself.

There was no avoiding it, I'd have to go through customs and get past the border guards. But how precisely was I going to smuggle a kitten across an international border?

That was the question.

At the height of my partying days back in Scotland, I'd had some experience slipping weed and alcohol into music festivals. I'd hidden stuff in my shoes, in my headband, in all sorts of places, with mixed success. I'd been busted a couple of times, but got away with a slap on the wrist. This was different.

Officials in this part of the world carried guns.

I sat at the side of the road, staring at my bike, hoping for a brainwave. I couldn't slip the cat into the panniers at the back. Apart from anything else, there was no room. They were stuffed full of my gear. For a moment I considered putting on the heavy jacket I had packed away. I could tuck the cat inside. But again, that was a daft idea. The chances of a wriggly, nervous kitten sitting there quietly were virtually zero. It would want to say hello to the border guard, guaranteed.

So my only real option was to zip the kitten up in the pouch at the front of the bike and hope the border authorities didn't notice it. That wasn't going to be easy. The wee thing hadn't stayed quiet before, so why would it do so now? But I had no choice. I had to take the risk.

I played with the kitten for a while in the hope I'd exhaust it. There were some long-stemmed daisies growing nearby. I grabbed a few and let the kitten chase the stringy stems around. It went crazy, running around in circles and springing up and down as if it was on an invisible trampoline. For a while I despaired. It wasn't slowing down at all; the cat was an insatiable ball of pure energy. The Duracell kitty. But then, as if by magic, after about twenty minutes its batteries ran out and it lay on some rocks next to me, as if ready to snooze again. It was time to make my move. 'Okay,' I said, steeling myself. 'Let's do this.'

I was encouraged to see a sudden flurry of traffic heading towards Montenegro. If I was lucky, maybe they'd still be there when I passed through. They might distract the guards, make them less interested in me. No such luck. When we reached the border about ten minutes later, there wasn't a single other vehicle in sight. It was only me. Or me and my stowaway cat, to be precise.

The crossing was a modern construction, a series of barriers and booths stretched out under a metal frame with a brick building and some offices attached. I pulled up alongside one of the booths, being careful to park the front of the bike past the window, so that it was out of the guard's eye-line. The kitten was still snoozing, but I was paranoid that it would wake up and start meowing. So I kept my sound system playing at a low level. The young customs official was behind a glass partition, which was a help. With luck, he wouldn't hear the cat, even if it did start making a noise. It would be drowned out by the gentle *thud, thud, thud* of my music.

The guy appeared totally bored. He flicked through my passport casually, not even bothering to check my photo or ask any questions. He then reached for his stamp and searched for a clean page to mark. I tried to remain calm and kept smiling, looking at him directly in case he tried to make eye contact. We were nearly through. But then out of the corner of my eye, I noticed the pouch. Its surface was rippling and, at the point where I'd left the zip open, the kitten was trying to push out its paw. It was also meowing. Loudly.

My heart nearly jumped out of my chest. Somehow, I managed not to swear, which was a feat for me. I forced myself to hold my nerve and kept looking towards the guard. For a moment all I could hear was meowing. There was no way he wouldn't pick up on it, I was convinced of it.

I'm no great believer in spirits and guardian angels. But one of them must have been watching over me at that moment, because it was then that a small truck suddenly appeared. It was a battered old vehicle with a noisy exhaust. The truck quickly drowned out the meowing – and almost everything else.

The official stamped the passport and handed it back with barely a glance of acknowledgment. I could only have been at the window for a minute, but it felt like an hour. I pushed off, not even daring to look back behind me. My elation was short-lived. We'd left Bosnia. But now we had to tackle the border into Montenegro. Leaving a country was one thing, entering quite another. This would be more challenging, I knew it.

Sure enough, the second border had more of a military presence. There were a couple of guys with guns walking around a large lorry that had been pulled over.

I cycled steadily and went through the same process again, placing the pouch as far from the window as possible. But this time I took extra precautions. As well as turning the music up a wee bit, I put my finger into the pouch every now and again to let the kitten play with it. A couple of times it dug sharply into my fingers, but I tried not to flinch. It wasn't easy. Its little teeth were like needles and they really stung.

This guard was a lot more attentive. He held the photo ID section up and looked at me. He stroked his chin as if to indicate my beard was a lot thicker than it was in my passport photo. I nodded and smiled. He didn't speak any English, so I wrapped my arms around myself as if to indicate it kept me warm. He just nodded.

The sound of his stamp punching into my passport was the most wonderful thing I'd heard that day. I climbed on the bike and pushed off past the barrier and down the road, feeling like the weight of the world had been lifted off me. I was ready to celebrate and get the kitten out of the pouch. But as I was about to pull over, I turned a bend to discover – to my horror – there was another checkpoint. This was one was smaller and less intimidating. But it still might catch me out. I approached it slowly, praying that I wasn't going to be third time unlucky.

Don't do anything stupid, Dean.

I was about to bring the bike to a stop again when a guard emerged from a small booth. He was on his phone and seemed preoccupied. He simply waved me through, barely offering me a second glance while he talked away. I gave the guy a nod and a thumbs up, then pushed on.

I was tempted to put on a sprint, but thought better of it. I didn't want him to think I was some kind of felon, rushing away from a crime scene, even though – technically speaking – that's precisely what I was.

3
Second Chances

A few miles on from the border, I came to a stretch of open countryside where some major roadworks were under way. The construction team had obviously taken the day off; there was no sign of anyone, and the diggers and tractors were parked up. I pulled off the road. My knee was complaining after the long climb. I also needed a breather, to calm my nerves and weigh things up after the morning's dramas.

I sat down on the tracks of one of the diggers and let the kitten explore. It was soon dashing around, getting excited about a patch of grass here and a pile of concrete kerbstones there. It didn't know what it was looking for, nor did it much care. The poor thing was having fun; it was certainly overdue some.

I took a couple of pictures of the cat then spent a few minutes on my phone, scrolling through a website that had a list of vets in Montenegro. The best one seemed to be in the coastal town of Budva, a few hours down the road. It was touch and go whether I'd get there in time tonight, but I decided it was worth a try.

Before hitting the road again, I decided to have something to eat, spreading some more pesto out for the kitten at the same time. For a few minutes I sat there, soaking up the winter sun, but also mulling over what had happened today so far. It had been an adrenalin rush, that was for sure.

I was distracted by the sound of a car engine. I turned and saw a battered, old silver Volkswagen Golf emerging on to the main road from a small lane that led to some fields. There was a young lad at the wheel. He couldn't have been more than eighteen or nineteen years old. He had a mate with him, and loud music was booming out of the window. They were laughing and waving and shouted something at me. As they disappeared down the road, I smiled to myself. It was as if I was watching a scene from my past. My dad used to have that exact same car, and my journey to this place may well have begun at the wheel of it, on a long and eventful night four years earlier.

It was the sort of stupid stunt that had been typical of my life for the past decade or so. Alongside me in the car that night, as was often the case when I was up to no good, was my long-time friend Ricky. We'd been partners in crime – lovable rogues, we liked to call ourselves – for ten years or so, since my early twenties. We regularly hung out, usually smoking weed and getting up to mischief. Ricky and I shared the same taste in music, had a similar outlook on life. We were both party animals – free spirits, I guess you might say. We didn't do things by the book.

That was certainly the case that night. We borrowed my dad's car without telling him our plans. We then drove about an hour and a half from Dunbar up to a field in Kinross, more than sixty miles away. But it wasn't any old field. It was going to host the big T in the Park music festival in a week or so. We went most years. It was one of the highlights of our summer, a long, sunny weekend watching some of the biggest bands around while smoking and drinking to our hearts' content.

Our crazy plan was that this year we'd bury some weed at a little spot of open land that we'd be able to recognise when we wandered into the festival later. We'd dig up our secret, personal stash and – hey presto – we'd be sorted for the three days of the festival. We thought we were geniuses. Except we weren't, of course; far from it.

We'd driven in the middle of the night to make extra sure we weren't spotted. The organisers hadn't started building the site yet, but we knew from experience where the perimeter fence and stage was likely to be. Having found the right place for our burial in the torchlight, we headed straight back down the motorway. As the only insured driver, I'd driven up and back again but had been working all day, so by the time we were half an hour or so from Dunbar, I was nearly falling asleep.

I remember closing one eye. The next thing I knew, we veered off the road and hit one of those raised platforms where police cars sometimes watch the traffic. The impact shot us into the road, where we hit the central barrier then started flipping over and over. We tumbled down a thirty-foot hill into a farmer's field. Everything seemed to happen in slow motion, as if we were in a movie. I can still recall how the airbags came up and hit us in the face, and being tossed around as if we were in a tumble dryer. Most of all, I remember us coming to a halt and Ricky and me sitting there upside down, with the roof of the car caved in to within centimetres of our faces. We hugged each other and then sat there, dazed, smeared in blood from a few small cuts. Shaking but, incredibly, otherwise unharmed and so glad to be alive.

Surviving a serious car crash is a life-altering experience. You feel like you've cheated death, been given a second chance at life. It was certainly a big turning point for me; it changed my perspective. It gave me a real motivation to do more, to experience more. I kept telling myself not to waste a single day. So when, early in 2018, Ricky first floated the idea of us doing some travelling, my ears pricked up straight away.

We were sitting out in the open air, doing what we did far too often – smoking weed. For some reason, Ricky started talking about touring around South America. It appealed to me on more than one level. A couple of years earlier, I'd been to Thailand with a girlfriend. I'd been fascinated, but also frustrated, as we passed through places on buses and in taxis. I wanted to know more. Who lived there? What was their life like? After returning home, I felt a need to experience the world up close and personal, not as a tourist.

There was another, much more personal reason for wanting to travel. I'd begun to realise that I had to get away from my life in Dunbar. The car accident may have been a catalyst, but it only added to a feeling that had been brewing for years.

People often wonder whether cycling around the world is a sign that I'm running away from something. There might be some truth in that. It wasn't that I needed to get away from my family. We had our ups and downs, but my mum, dad, sister and I were close and I still lived at home with them and my gran. I wasn't bursting to set up a new home on the other side of the world, either. Again, I enjoyed the town and community where I'd grown up; Dunbar was a great place, filled with great people. No, if I was trying to break free, to run away from anything, it was my old self. And the meaningless, routine life that I'd boxed myself into.

I was a good person at heart, I knew it, but I had always been mischievous, the eternal class clown. There had been times when it had got me into hot water, especially when I had too much to drink, which was fairly often back then. I'd been fined by the police a couple of times and had got into the odd fight here and there. I was normally laid back, but I could get fiery when I had too much booze. Drink could be the devil in that way. As I headed towards my thirtieth birthday, I sensed it was time to do something different. It wasn't that I felt I was going in the wrong direction; it was more like I was going round in circles.

My parents and sister were all in the caring professions. My mum was a charge nurse in the NHS, while my dad had worked in mental health and was now a foster carer. But ever the rebel, I'd pushed against all that. It didn't appeal. At school, probably encouraged by my grandad, who'd served in the army, I fancied enlisting and started basic training with the Royal Electrical and Mechanical Engineers. But after a few months I decided against that too and dropped out. Afterwards I had a couple of long-term manual jobs around Dunbar, working as a handyman on a farm and then as a welder at a fish-food factory.

I had always been good at fixing and building stuff, but I hadn't built much of a life for myself. While some of my school friends had settled into careers, bought houses and started families, I hadn't settled down at all. I felt that by hitting the road and spending time in another part of the world, I might somehow find myself. Or, at least, find a way of being myself. Someone said to me once that I hit the road to find a road. That seemed about right.

In the weeks that followed that first conversation with Ricky, I became more and more excited about the idea. I was a keen cyclist, as was Ricky. I had the idea that we should cycle to South America the long way round, via Europe and Asia. I felt it was a once-in-a-lifetime opportunity for us, something that would feel like an achievement when we looked back on it years later.

'Imagine telling your grandkids that you cycled around the world when you were younger,' I said to Ricky, trying to sell him on the idea one night in the pub.

He'd not taken much persuasion.

I was a little worried about telling my parents, but to my surprise, they seemed glad that I'd decided to do something with my life. The truth was, they'd been even more concerned than I was that I'd been heading down the wrong road. They both thought an adventure would be good for me. My dad's view was that it would be 'character building'. Their approval gave me the extra encouragement I needed.

Both Ricky and I knew we'd need money to make the trip happen, so we agreed to save up. Neither of us were afraid of hard work, so we grafted for six months. While Ricky worked in a cement factory, I got a job building a railway line in a theme park in Glasgow. We took on extra jobs as well, working in bars and doing bits of manual labour. At one point we had five jobs between us and were working an average of eighty-two hours a week. By the autumn of 2018, we'd put a few thousand pounds each to one side. We'd also started mapping out a route, first to mainland Europe, then down through France, Switzerland and Italy to the Balkans, and on to Greece. I'd started putting my gear together as well.

I was determined to get myself the best possible bike and splashed out on a top-of-the-range, off-white, Trek 920 tourer with drop handlebars and specialist, mountain bike wheels. It cost me nearly two thousand pounds, but the minute I got it out of the box, I knew it was worth every penny. I particularly liked the fact it was a lightweight model and weighed under 13kgs, or 28 pounds, without any extra baggage on board.

After a test ride or two, I decided to make a couple of modifications, adding some bigger, stronger pedals and a new saddle. We were planning to cycle a long way, after all.

I'd soon fallen in love with the bike. I would come back from a ride and park it in my parents' yard and sometimes stare at it in wonder. It was a thing of sheer beauty. I was so smitten, I even gave it a name: Eilidh, the Gaelic version of Helen, meaning 'sun' or 'radiant one'.

As I started gathering the rest of the kit I'd need for the trip, I also bought a one-wheeled trailer that I could fix to the back for extra storage.

Ricky, by contrast, was content with my old bike, a mud-spattered old Trek tourer that I'd ridden to death over the past few years. It was a good bike and Ricky was determined to ignore all those who told him to get a new one like me. 'If it can travel one mile, it can travel twenty thousand miles,' he argued.

He did it up with new tyres, a fancy new 'Brooks' saddle, and other bits and pieces. It seemed to run well and survived our regular practice rides without any problems, even making it to the top of the 600-feet-high North Berwick Law, a steep mountain climb about an hour away. We spent a night up there in preparation for the months and maybe years of wild camping that lay ahead.

And so it was that – in September 2018 – we set off with a very rough plan.

It fell apart almost instantly.

We got off to the worst possible start, mainly because we behaved like the same pair of idiots that we'd been back in Scotland. We'd planned to sail from Newcastle across to Amsterdam after cycling down the north-east coast of the UK from Scotland into England. But we had such a big party on the night before we left that we didn't manage to set off until 5 p.m. the following day. We'd still been drunk; it set the tone. From then on, we treated the entire trip down as a glorified pub crawl. It should have been no surprise when we quickly found ourselves behind schedule for the ship on which we were booked to sail.

The chickens came home to roost when Ricky somehow managed to lose a tooth and we had to go to an emergency dentist in the little town of Alnwick, about thirty miles from Newcastle. I was sitting in the waiting room going through our paperwork. I thought our ship was sailing the following evening. Except it wasn't. It was that night. In fact, it was now five o'clock in the afternoon and it was leaving in an hour and a half, at six-thirty. There was no way on earth we'd make that journey. We'd missed it. We were stranded for days.

When the ferry company managed to squeeze us on to a boat, my mum and dad came down to wave us off. There weren't too many tears; we all knew I was doing the right thing, but my dad did give me a little memento to remember them by. He is from Newcastle and a passionate Newcastle United football fan and had bought me a small pin badge with the club's famous crest of seahorses and black-and-white stripes for luck. I clipped it to my rucksack then headed for the boat.

Sailing across to Amsterdam, Ricky and I pledged to take things more seriously, but nothing changed. We were soon back to our bad old ways. The first thing we did on arrival in Holland was to go to a weekend-long rave.

Every now and again we'd give each other a pep talk. 'We can't carry on like this,' we'd say to each other. But we couldn't stop ourselves. We were such a bad influence on each other.

The more we travelled, the more obvious it became that we also wanted different things. We passed through Belgium and into France, heading into Paris, which wasn't the plan I had in my head at all. I wasn't a fan of big cities. I wanted to see open roads and countryside, experience different wild environments and meet interesting people – not eat overpriced food and battle with tourists to walk a hundred yards down a boulevard. Ricky, meanwhile, wanted to be with his girlfriend back in Scotland, rather than with me. In hindsight, it was a flawed plan from the beginning.

Our time together wasn't without its fun moments. It was full of them, truth be told. Heading through Switzerland, we took the mountain road in the Furka Pass past the famous old Hotel Belvédère, best remembered from the James Bond movie *Goldfinger*. It was abandoned and boarded up, but we found a broken window at the back and let ourselves in. We had a suite each to ourselves.

But I also got a glimpse of what it might be like riding solo, when Ricky broke off to see his girlfriend and I cycled through part of France on my own. I loved being on the open road, wild camping wherever and whenever the mood took me.

Once we'd been reunited and hit Italy, there was all sorts of new trouble. At one point I had my passport stolen and had to head back to Glasgow for a new one. At another, Ricky had his bike stolen. Miraculously he got it back, but minus its panniers containing a lot of his stuff. After that, his mood turned more downbeat. It wasn't helped by the fact that he was running out of money. Everything came to a head when we cut through Croatia and down into Bosnia, reaching the town of Mostar.

Ricky had been invited to a stag party up in Hungary, in Budapest. He was cycling up to the party – and calling it a day after that. He wanted me to go with him. I would then be free to cycle back down and carry on my journey alone. But I was not keen on the idea. The weather forecast was terrible and I had my eyes set on heading south to Montenegro and Albania, and after that Greece. I said no.

We didn't have an exchange of words; we simply decided to head in different directions. At the end of our final day together, I stayed in a hostel in Mostar while he set off back towards Budapest. That was that. No hug, no handshake, no goodbye.

For a while I was upset. The original plan hadn't worked out. I felt as if I'd screwed up my big chance. Maybe my last chance. But slowly I began to see that it had been the right thing to do. The only thing, if I was honest with myself. If I was going to make this adventure work, I'd have to do it in my own way, and in my own time.

Sitting on the roadside in Montenegro almost a fortnight later, the decision to go our separate ways seemed wiser than ever. This morning's adventure proved the point perfectly. Would we even have noticed the kitten? Would we have taken it on board with us? If we had, would we have made it through the border with it? I could never know, of course, but I had my doubts.

None of this reflected badly on Ricky. The opposite, in fact. I realised that I owed Ricky a massive debt; he'd shown me exactly how *not* to tackle this journey. I could now see a path ahead.

The kitten scampered back into my lap and snapped me out of my thoughts. It curled up alongside me, breathing a little erratically as if it was out of puff. It was probably exhausted after its time being exposed and alone on the mountain. I gave it a reassuring stroke and it pressed itself closer to me. It felt good to know it was out of danger, that it seemed to feel safe and secure in my company.

Above all, I was pleased that I'd been able to give it a second chance. I had no idea what lay in store for us, but maybe my scrawny new companion would help me make the most of mine as well.

4
Room Mates

By mid-afternoon, the weather had taken a turn for the worse. The clear, blue winter skies of earlier had been replaced by iron-grey clouds that were threatening heavy rain. Where the road was most exposed, I had to lean into stiff headwinds that made cycling tough. They slowed me down and my plan now was to reach Budva by early evening; too late to see the vet today, but before the heavens opened.

My passenger was still coiled snugly around my neck, more comfortable than ever. The kitten was no longer sleeping; I could feel its head twitching from side to side. Maybe it was enjoying the view, which would have been understandable. Even in the worsening light, the Montenegrin coast was a spectacular landscape of mountains and lakes, ancient churches, and pretty villages with red-tiled roofs.

I still couldn't believe that I'd got away with smuggling the little creature across the border. Deep down, I was still half expecting to be flagged down by the police or some other local authorities at any moment. I felt sure that anyone watching me closely would have guessed I was up to no good. I might as well have 'International Cat Smuggler' emblazoned on my T-shirt.

In late afternoon we reached the car ferry that crosses the lake at the beautiful town of Kotor. Approaching the boarding ramp, I saw a uniformed ticket inspector walking along the lines of cars, talking to the drivers of each as he went. I panicked. Without really thinking, I squeezed the kitten back into the pouch. Its meows of protest were drowned out by the heavy chugging of the ferryboat's motors. The inspector barely registered me as I handed him the right money and collected my ticket.

It was only when we were halfway across the lake crossing that it struck me. What the hell was I worrying about? Why would a ferry inspector care about a kitten? Why would anyone on the boat care, in fact?

Sure enough, when I let the kitten clamber back on my shoulders, people were charmed by it. One little boy began jumping up and down excitably inside his dad's car, pointing at us. Other drivers and passengers nodded and smiled. I felt slightly self-conscious at first, but when I thought about it, I couldn't blame them for staring. We were an unusual sight; a big, bearded tattooed bloke on a bike, with a kitten sitting on his shoulder like Long John Silver's parrot. We were going to draw attention. But, more importantly, their reaction was proof that I didn't need to keep worrying. As far as the world was concerned, the kitten and I were travelling companions. We may have been an odd couple, but we were a couple, nevertheless.

By the time Budva loomed into view, the sun had disappeared over the mountains to the west and the light was fading fast. It was a relief to reach my destination; I'd been cycling for almost seven hours in all and had racked up nearly sixty miles. My injured knee was beginning to scream for mercy. So I was relieved to find a small campsite set in some gardens near the town's beach. While I quickly rigged up my tent, I left the kitten to run free and explore its new surroundings. It was still cautious and didn't stray far. If there was a sudden noise or it found something unexpected, it came scampering back to me. It had obviously started to place its trust in me, which made me smile. I could feel the bond between us growing.

On the way through town, I'd popped into a local shop and got some food for us both, pasta for me and pesto for the kitten. There was hardly any proper cat food on the shelves, and it seemed to like pesto. I cooked up the pasta on my little stove then ate dinner alongside my new friend outdoors, overlooking the sea, not that you could see much. The horizon was a grey blur now and the stiffening sea breezes were carrying the first flecks of rain. I was about to pick up the kitten and take it inside the tent when it started meowing loudly. It was a different sound to any I'd heard so far. More intense. It was also looking at me as if trying to tell me something. It had been fed and watered, so I decided it probably now wanted to go to the toilet. I picked it up and carried it to a wall overlooking the sea. As expected, it jumped down onto the sand and scurried out of sight; in search, I assumed, of a little privacy.

For a while, I simply took in the view. The beach stretched for half a mile along the coast and was deserted apart from a man playing with his dog. It must have been a few minutes later, as the man and his dog approached the wall, that I began to wonder where the kitten was. Is that it, I asked myself. Has it run off without even a thank you? I hoped not. But I needn't have worried. I was about to jump onto the sand when it came shooting back to me like some guided missile, leaping up onto the wall in a single bound. Maybe it had picked up the whiff of the dog approaching, or perhaps it simply wanted to be back with me. I was happy it had returned, either way. I would have worried all night if it had gone missing.

We got under cover as dark descended and the storm arrived. I was soon listening to the whistling of the wind and the *pitter patter* of the rain on the roof of my tent. As I lay there on my phone, I spent the evening doing what I normally did – watching videos and updating my Instagram.

Ricky and I had set up the page for our trip before leaving Dunbar, partly as a way for everyone back in Scotland to keep track of us, but also as a sort of scrapbook documenting our progress. We'd picked up a number of followers during our travels through northern Europe and now I was chuffed to have close to a couple of thousand. Today's chance encounter seemed like something that should definitely be included, so I shared the video I had taken of the little kitten on the roadside, along with a picture of it sitting safely in the pouch of the bike. People really liked them both.

I enjoyed being alone in the tent, but it was different tonight. I no longer had my space to myself. Like any new roommates, it took us a while to get used to each other. At first, the kitten was restless and slightly wheezy again after all its exercise. It couldn't settle in one spot. It tried lying by my feet, then at the back of my neck, fidgeting all the time. For a while it draped itself across my thighs, but eventually it found a comfortable spot across my chest, near my face, where it lay curled up in a ball so tight it could have tied itself in a knot. Its breathing was uneven on occasion, but it seemed comfortable enough. It had soon slipped into a deep sleep.

I was exhausted from the day's cycle, so wasn't far behind it in drifting off. The sound of the wind and rain growing stronger outside helped me on my way. It must have been two or three in the morning when I stirred in my sleeping bag. I still felt groggy and disoriented in the dark. The kitten was sleeping across my feet again, I noticed, but that wasn't what had woken me. There was something else, a very nasty smell.

It took me a moment to work it out, but I realised it was coming from somewhere inside my open sleeping bag. I shone my torch and saw there was something smeared on the bottom. It had got on to my legs as well. It was yellowy and oily. And it stank. It was obvious what it was. All the pesto the kitten had eaten had gone straight through it. I didn't know whether to laugh or cry. What a great way to repay your new roommate, I said to myself. After all I'd done for it.

By now the wind was howling and it was pouring with rain, so I did what I could to clear the mess from inside the cramped tent. But the smell lingered all night. It was a relief when I woke at daylight. The wind and rain had eased and I went to the standpipe outside to give the bag a good wash. The kitten inched its way out, looking as if butter wouldn't melt in its mouth, and I laughed. I knew this new relationship was going to be a steep learning curve.

And here was my first lesson.

'No more pesto for you my friend,' I said, as I hung the sleeping bag on a branch of a nearby tree to dry in the gentle morning breeze.

I rang the clinic first thing. They could see us that morning, so I headed into town after breakfast with the kitten perched on my shoulder. As we explored the old town and I took a few photos, we drew more than a few curious looks. But most pointed and smiled. A couple of children even came up and asked to stroke my companion. I was happy to let them. The kitten seemed to enjoy the attention.

The vet surgery was a modern, well-equipped clinic up a hill in the old town. The vet, a bearded, bespectacled guy, spoke very good English, which was a relief as my Montenegrin was non-existent. He began by examining the kitten thoroughly, looking at its teeth and eyes and running his hand up and down its ribs and back. I thought it might make a nice photo, but he scowled when I pointed my phone at him and the kitten.

'If you keep doing that, you can wait outside,' he said sternly.

I put the phone in my pocket. It wasn't worth it for a couple of snaps for my Instagram.

'A little skinny,' he said, after a few moments. 'She just needs a good meal.'

'She?' I asked.

'Yes, female definitely. About seven weeks old.'

I mentioned the little wheeziness that came on after she had been running around. He stuck a stethoscope on her and had a listen to her lungs.

'Where did you find her?' he asked.

'Abandoned, by the road in the mountains.'

He shook his head mournfully.

'Happens a lot, unfortunately,' he said. 'People just throw them out of their car windows. She's probably been outdoors in the cold. Her lungs are a bit weak, but should get stronger as she grows. Keep an eye on it.'

To my relief, there was no sign of a microchip. Even if there had been, there was no way I was going to return her. Whoever had dumped her didn't deserve to own any animal.

'So, what are you going to do with her?' the vet enquired, as he applied a worming treatment to her skin.

It was the first time anyone had asked me that question. But if I was honest with myself, I'd known the answer since our unlikely meeting back in the Bosnian mountains.

'I'm going to keep her,' I replied. 'Thought I'd take her around the world with me.'

He looked a little startled, but soon reached into a drawer and produced a form.

'You'll need a passport then. Border guards don't like undocumented animals.'

I wasn't going to have anything bad on my conscience, so I grabbed the form. 'What do I need to do?'

'We will need to get her microchipped and she will need to have her jabs. I can give her one today and another one in a week. I'll do the microchip then as well. We can then issue the passport.'

'Great,' I said.

I had no problem spending a week or more in Budva. It was a pretty place and I was sure that once the cat and I had settled into a routine, we'd be very comfortable at the campsite. The weather forecast for the next few days was rotten too, with tons more rain coming in. It would give me a chance to get to know the kitten better and to buy the extra gear I'd need if I was going to take on board a permanent new passenger.

The vet gave the kitten her first jab. She winced a little as he inserted the needle, but I held onto her paw and she'd soon forgotten about it.

'So, you'll also need a name,' he said, handing me a bill and a card machine for me to pay. 'For the passport.'

That threw me. I hadn't given it any thought.

'Can I tell you when I come back next week?' I answered, handing him the card and then punching in my pin number.

'Of course.'

I'd spotted a pet store in the town, and on the way back to the campsite I popped in to buy a few basics. In addition to a couple of plastic bowls for her to eat and drink from, I got her a toy, a little mouse on a string. I also picked up a harness. She'd nearly fallen off the bike a couple of times already, and I was conscious that she was still young and naive. She could bolt out in front of traffic, or jump off from a height and hurt herself. The harness would keep her safe.

Finally, I bought a carrying case. There had been two to choose from in the store. One was an enclosed black bag that looked a little too claustrophobic. So instead, I chose a colourful one with cat designs and a wee window on the side for her to watch the world go by as we cycled.

Back at the camp, I fitted the case on the back of the bike using some strong rubber straps. It fitted neatly enough. Unfortunately, the same couldn't be said for the harness. It had been the smallest size the pet shop had in stock, but even then, when I tried to put it on, it was too big. She was such a tiny thing. It wouldn't keep her safe, she'd slip her head out of it easily. I was about to discard it when I had a brainwave.

I popped into another shop nearby and got some superglue.

Back at the camp, I cut a section out of the collar and put it around her neck so that it fitted. I then superglued the collar back together again.

'There you go,' I said, admiring my handiwork.

Later in the day I took her for a walk at the end of the lead. She wasn't happy. She kept pulling on it, mewling and fiddling with her neck at the same time. It was only when we got back to the camp and I tried to take it off that I saw the problem. Part of the collar had got glued to her.

'Oops, I'm a bad cat dad,' I said.

Thankfully, once the collar was off, I had a pair of scissors and I started trying to take off the little snippets of glued-up fur. It wasn't easy; the kitten wriggled and darted away from me constantly. It took me half an hour to remove all the sticky stuff properly. Late that afternoon the sun made a fleeting appearance, so I tried to make it up to the kitten by taking her for a ride along the coast. We soon found a beautiful cove with an abandoned building overlooking a stunning little beach. I parked the bike and set off to explore.

We had the beach to ourselves. The kitten ran around ahead of me, sniffing at the driftwood and other bits of flotsam that had washed ashore. She seemed in her element, as was I. The sun was soon dipping behind the mountains once more. I sat on some rocks as the cat jumped around. She was a brave wee thing and at one point leapt what looked like ten feet from one boulder to another. Afterwards she stood on a large rock that was jutting out to sea, gazing out over the coastline like a proud, little lioness. It was as if a lightbulb had been switched on in my head.

One of my favourite films as a kid was *The Lion King*. And my sister's favourite character was Simba's childhood friend Nala, who grows up to become his wife. She was a feisty and courageous little character too, I seemed to remember.

I had a decent signal on my phone, so I looked it up. Scrolling through some stuff on the web, I also saw that Nala apparently means 'gift' in Swahili. That clinched it. We'd only been together a day, but the little kitten already felt like a gift. Rather a precious one, if I was honest.

'That's it. Decided,' I said, playfully ruffling the back of her neck. 'Your name's Nala.'

5
Riders on the Storm

The next few days drifted by in a haze of grey that reminded me of the *dreich* Decembers back in Scotland. It rained almost constantly. So, apart from the odd visit to the shops and toilet trips to the beach with Nala, I was mostly confined to quarters, inside my tiny tent.

Boredom wasn't an issue, though. Nala was a non-stop entertainer; she could spend hours play-fighting with me or chasing the little mouse toy I'd bought her. At night, if I shone my torch at the skin of the tent, she'd leap around energetically trying to catch the little dot of light. No matter how often she tried and failed, she never seemed to get disheartened or disappointed. I never tired of watching her antics.

She was really affectionate, too. She had taken to nuzzling close to me and rubbing her nose against my forehead as I updated our story on Instagram, or watched Netflix on my phone. All in all, she was brilliant company.

As the week wore on, looking after her became much easier. There had been no repeats of the pesto poo incident on the first night, thank goodness, and by now I'd found a shop in Budva that had some decent cat food. We often ate together, me munching on takeaway chips and pasta, while she licked her little bowl so clean she could see her reflection in it. Beyond nagging me for her meals and toilet trips, she was really low maintenance. As long as I was near, she seemed to feel safe and secure. And, unlike a human companion, I wasn't expecting her to have any opinions about where we went and when. She'd leave that to me. I had a feeling she was going to turn into the perfect travelling partner.

How good she'd be once we hit the road again remained to be seen, of course. So, with me itching to get back on the bike, I returned to the vet six days after our first visit. He gave her a second jab and microchipped her, slipping the little chip into a small incision at the back of her neck. He checked her lungs out, too. They weren't any worse than before, which was good news given how damp it had been. He then filled in the paperwork for the chip, as well as her 'passport'. It was in a blue wallet and was technically called an International Certificate of Vaccination and Veterinary Health Record. It was in English and Montenegrin, which was helpful.

Seeing my name and home address alongside Nala's name felt good. It was official, she was mine. Well, as far as the authorities were concerned, in any case. I wasn't convinced anyone could ever 'own' an animal, especially one as independent as a kitten. To me Nala was, and always would be, a free spirit. Like me.

The vet handed me a bill that, like before, was in the local Montenegrin currency. I once again punched my pin into the machine.

'So can I cross into Albania now?' I said, as I waited for the payment to go through.

He seemed a little surprised. 'No, she will need a rabies jab to get through the border.'

I expected that was simply a matter of booking another appointment.

'Okay, I need to move on. Can we do that in the next few days?'

He looked at me as if I was stupid.

'No. Only when she is at least three months old,' he said, glancing at his paperwork. 'I have put her date of birth as the second of October. So, we could make an appointment for you sometime late in January or early February.'

My heart sank. I had a rough timetable in my head and I wanted to be in Greece by January, if not before the New Year. This was really going to slow me down.

'Is there no way she can have it earlier?' I asked.

He shook his head disapprovingly. 'No. You will have to wait.'

I spent the rest of that day and night in the tent, really torn as to what to do. I looked at the map. It was a couple of days or maybe fewer to the Albanian border. After that, it was probably a week or two's cycling down to Greece. With Christmas now a week or so away, I could easily make the Greek border by early January. Why not make a dash for it? It was a familiar feeling; my mind kept flipping one way, then the other. Who said I had to be in Greece by January? What was wrong with staying here?

On the other hand, while I was grateful for the help this vet had given us, he wasn't the only one in the world. There would be others along the way. Why not make more progress and take Nala to another vet? Let them deal with the next set of jabs, whatever they may be? I'm normally a good sleeper, but that night I tossed and turned more than usual. By the time I nodded off, though, I'd come to a decision of sorts.

I woke up the next morning with Nala's face next to mine. She was licking my forehead and breathing gently on my face. As I blinked myself awake, she added a couple of plaintive meows for good measure. This was another lesson I learned quickly from her. She didn't hold back when it came to feeding time. This was her way of saying: *Come on, mate, where's my breakfast?*

I rolled out of my sleeping bag and gave her something to eat. I then stuck my head out of the tent. To my relief, there had been a break in the weather. It wasn't exactly tropical sunshine, but the rain seemed to have disappeared for now. For the first time in days, I could see the coastline stretching for miles to the south. I took it as an omen, a sign that I should head for the Albanian border. I packed up camp, placed Nala in her colourful new carrying case at the back of the bike, and set off mid-morning.

It wasn't long before I had to make my first stop. Within minutes of us getting under way, Nala started making a real racket, meowing so loudly that at times I thought it was a police siren going off behind me. At first, I put it down to her not being used to her new carrier. She would settle down, I told myself. But her meows got fiercer and louder. When I looked back, I could also see that she was putting pressure on the case. She was obviously trying to break out.

I pulled over and took her out. She immediately jumped onto my shoulder. I let her sit there for a while so that she could calm down. But when I got the chance, I transferred her to the pouch at the front of the bike again. She curled up inside the pouch for the first part of the journey, but after a while started poking her head out as if to check on our progress. She seemed much happier there. It gave me the encouragement I needed to press on and try to get some good mileage under my belt.

The weather forecast I'd looked at online had talked about stiffening winds and potential storms, but at first the conditions were fairly good; the skies were grey and there was only a mild sea breeze. I thought I might get away with it, but I'd not travelled very far when all that changed.

I was on a long stretch of open road. According to the map on my phone, it stretched for many miles south and should – in decent weather, at least – have speeded up my journey. It didn't work out that way.

The first thing I noticed was the wind. It wouldn't have been so bad if it had been at my back, or even hitting me from the side. But it was blowing straight at me and was soon making cycling really difficult. I was a strong cyclist and had pushed through some challenging weather before. But I was suddenly straining, even in the lowest of gears. It was as if I was being pushed backwards. To make matters worse, the occasional gusts were so strong they were buffeting me. One nearly knocked me flying off my bike as a giant lorry went past. If I'd toppled over, I could easily have fallen into its path, and there would not have been much left of me or Nala if I had.

I'd been on the road for about half an hour when I noticed the skies darkening. The grey slab of cloud of earlier was replaced by a menacing block of charcoal black. In the distance I could hear the loud rumbling of thunder and the odd flash of lightning. There was a storm brewing and I was headed right into it. Soon it was raining steadily. I stopped to check on Nala. When the winds had first started, I had wrapped her up in a towel so that only her head was peeping out. But she had quickly retreated inside the pouch, where she was now curled up in a ball. I wished I could climb in and join her. I zipped up the pouch completely to make sure she stayed dry. I didn't want to make her wheezy again, and I definitely didn't want her catching a cold, or worse.

From that point onwards, it was as if the storm gods had decided to turn the dial up to ten and beyond. On top of the strengthening winds there was now torrential rain, which was being hurled at me with such force that it hurt. I had only my shorts on and my legs were soon red raw from the force of the raindrops. It was absolutely miserable. I could barely see ahead of me.

I hit rock bottom when I started riding up an endlessly long hill. I was still cycling directly into the wind, but it now felt like it was blowing at gale force. Once again, I was being shaken by gusts, but this time they were knocking me clean off my seat. I had no option but to get off and walk. Even that was hellish work. I was leaning into the wind and pushing as hard as I could, while fighting to keep the bike from being blown sideways. At one point, I thought, *That's it, I've had enough. I can't go on.*

For a couple of minutes, I stuck my thumb out at passing traffic. But I doubted they could even see me, let alone stop safely on the exposed road. I soon gave that up. I had no choice but to press on alone. I must have travelled a dozen or so miles before I saw a sign for the town of Bar ahead. I decided to cut my losses; I wasn't going to make the Albanian border in this weather. It was also too risky; the conditions were, if anything, getting worse rather than better. My only consolation was that whenever I looked at Nala, she was still curled up asleep.

I rolled off the main road and headed towards Bar midway through the afternoon. A journey that should have taken two hours at most had taken five. By the time I arrived in the town I was exhausted. I was also soaked to the bone. I pulled up and booked into a hotel for the night.

I don't think I've ever been so grateful to feel the warmth of a room. I gave Nala a rub down with a towel then changed out of my wet clothes and had a warm shower. It was bliss.

I gave my wet clothes a wash and stuck them on the radiator to dry. That night we snuggled up while the rain and wind continued to rage outside. Nala was more restless than usual and I heard her coughing ever so faintly a couple of times. I felt awful. Why had I taken her out in that weather?

The rain was still falling the next morning. It wasn't as torrential as the previous day, but I wasn't going to risk it. I'd stay put for the morning and maybe rattle off a few more miles late in the afternoon; it would give us both more time to recover, to lift our mood. That proved to be wishful thinking.

Nala loved play-fighting with me. I'd wave my hand at her, coaxing her to jump and try to nip at my fingers, then pull away before she caught me. It didn't always work. Sometimes she succeeded in clamping her teeth around one of my fingers. It wasn't easy removing them, either, or the little indentations her teeth left behind. Other times our play would escalate into a proper scrap – a fake one, of course. She could be really energetic and rumbustious and loved it when I picked her up and let her drop gently on to a mattress or a table.

That night, I felt so bad about dragging her through the storm that I decided to indulge her. We were soon having huge fun wrestling with each other next to the bedside table. She was hyper-excited at this point and decided she was going to jump onto the table and clear everything that was lying on its surface. That wouldn't have been a problem normally, but I'd left my phone there charging. It took me a split second to realise my mistake.

If it had been a scene in a movie, I'd have been flying through the air in slow motion, shouting, with my arms outstretched in desperation. I would have seen my phone teetering on the edge of the table, and then – as my fingers were about to reach it – fall on to the hard stone floor with a loud *crack*. I knew it was broken. The screen was smashed and had turned black. I pressed the restart button, but it wasn't responding. I was gutted.

How could I have been so stupid?

It wasn't Nala's fault, it was mine. I shouldn't have left it there. I was an idiot.

I must have looked like I had the worries of the world on my shoulders when I walked into the lobby.

'Everything okay?' the hotel owner asked.

I held up my phone.

'Just smashed it to pieces.'

He held the palm of his hand up, as if asking me to hold on.

'Wait a moment. I might be able to help,' he said, picking up his own phone. Twenty minutes later, I was in a nearby phone repair shop with the owner's son.

'Broken screen and LCD,' he said. 'Leave it with me maybe two hours.' By that evening, I had my phone back. I'd been relieved of a couple of hundred pounds, but it could have been much worse. I still had six or seven hundred pounds left in the bank the last time I checked.

'Guess I've learned another lesson,' I said to Nala back at the hotel. 'Never play with you when my phone's around.'

We set off for the border midway through the following morning. Maybe it was because I'd had more time to think about it, but I felt more anxious than when I'd left Bosnia. I knew that Albania, until not too long ago, had been a Communist country and cut off from the rest of Europe. It was a lot more open to visitors now, but I still expected security at the border to be stricter.

I ran through the same drill I'd gone through back at the Montenegrin border. A few miles before the checkpoint, I stopped and got Nala out of her pouch. I sensed that the rough day we'd experienced had taken a lot out of her, too. Within about ten minutes, she was willingly curling herself up in the pouch again. When I zipped it closed, there wasn't a peep of protest.

The Albanian border turned out to be every bit as intimidating as I'd feared. The crossing looked like a miniature military camp. As well as austere-looking barriers and booths, there seemed to be barracks and some sort of headquarters behind it. There was a long line of cars backed up and there were guys in uniform with guns walking up and down. Some had mirrors to look underneath vehicles. But, as Nala and I eased our way towards the back of a queue, it was a pair of other guards that set me worrying. They were holding long leads attached to sniffer dogs.

My mind was soon running riot. I started imagining all sorts of doomsday scenarios. I knew the dogs were trained to sniff drugs or explosives. But there was no way they'd miss the unmistakable whiff of a little cat. The instant they did, they'd come racing over to us. And what about Nala? The smell and sound of the dogs might send her into a spin too. God knows what would happen then. These guys looked much more serious than anything I'd seen so far.

I told myself to stay calm. I had documents that proved she was mine. They wouldn't be able to confiscate her.

Fortunately, those guardian angels of mine must have been

floating around again. Within a couple of minutes of me joining the queue of cars, a new, shorter line opened up and I was waved into it. It put us further away from the sniffer dogs, whose handlers had been hauled over to another queue where they were checking a couple of big trucks.

I wasted no time in pulling up at the window and handed over my passport. The guy there asked me a couple of questions. When I told him I was heading to Greece and aimed to cycle around the world, he looked nonplussed. He shook his head and gave me a look, as if to say: *You're a nutcase, mate.* He stamped my passport and waved me on. It took less than thirty seconds. I breathed a huge sigh of relief but, as I cycled past the military barracks and into Albania, I didn't feel the same sense of elation as when I'd crossed into Montenegro. I felt like I'd been lucky – again. How many times would I get away with this?

I decided against riding much further that day. I didn't feel up to it, I felt tired and a bit out of sorts. So I headed for Shkodra, the first major city inside Albania, where I booked into a back-packers' hostel. I felt at home straight away. It turned out the owner had a couple of rescue dogs, which she kept out the back of the building, thankfully. She made a great fuss of Nala as we checked in, and gave us a lovely room. There was another guy there, a lanky, very talkative young Serbian called Bogdan. He was, like me, travelling through Albania, but backpacking on foot.

Nala and I spent part of the afternoon exploring the city. The quaint, cafe-lined streets of the oldest neighbourhoods were fairly quiet. It felt as if we were the only tourists there, which was hardly surprising given the time of the year. I took some photos and sent my drone up from the castle overlooking the city to film a little video. Back at the hostel, we relaxed by the fire in the communal area. I thought I could hear a hint of Nala's wheeziness when I put my head next to her chest. I was probably being paranoid, but I wrapped her up extra warmly in a blanket, just in case. She was soon snoozing.

While she slept, I chatted with Bogdan. He spoke very good

English and was a mine of information on where to go – and not to go – here in Albania. He recommended a few places for me to visit on the trip down south to Greece. I also spent some time catching up with stuff online.

At the top of my Instagram page, I'd added a row of flags that marked each country I'd visited. Albania was my tenth, if I included England. It made me feel as if my journey around the world was getting somewhere. But, scanning through the most recent photos and comments, I saw how dramatically the trip had changed for me this past couple of weeks.

Since splitting up with Ricky, the page had been called *1bike-1world*, a catchy little title that I felt summed things up well. One man going around the globe on his bike. Except, it was no longer accurate. It was now one man and his cat.

Since finding Nala, I'd posted a series of videos and photos of her. People back home in Scotland, where most of my now two-thousand-plus followers were still concentrated, seemed to appreciate them. They absolutely loved a picture I'd taken of her on the ancient walls in Budva. They also said lots of nice things about me rescuing her, especially when I posted the first video of me finding Nala.

This evening, I found myself reading and re-reading the comments. For some reason, it felt reassuring. Was I still doubting what I'd done? Did I simply need cheering up? It set me thinking and I found myself assessing my new situation in a way I'd not done before.

Our following had swollen by a few hundred since I'd found Nala. I'd noticed several newcomers were from overseas, some in the US. All this love and affection for her – and to a lesser extent, me. Did I deserve it? I wasn't too sure. Taking Nala through the eye of a storm had certainly been a big gamble. Especially with the question mark over her lungs. So had skipping out of Montenegro without the rabies vaccine and then smuggling her through another border today.

I shook my head quietly. No. There was no question about it, I'd been a bit reckless. What danger had I put her in? What might have happened to her if we'd been caught? I had been pushing my luck. The more I thought about it, the clearer it became what was bothering me, and what I needed to do. I had to take this more seriously.

Of course, we'd have to take the odd risk. If we didn't, we'd never get anywhere. After all, we were going around the world on a bicycle, not in a limousine or on a Lear jet. We were going to get caught out by the weather, by bureaucracy, and by my own honest mistakes every now and again. That was a fact of life. But I definitely needed to do better, to think about things more carefully at times.

The new reality was there, at the top of my Instagram page. 1bike1world. The truth was that the important number was no longer one, but two. It was about me – and Nala. The two of us. I had a responsibility to her as well as myself now. I had to start living up to it.

6
Resolutions

I spent the following morning working out my next steps, thinking not only about what was best for me, but for Nala as well. With the weather turning steadily colder, I wanted to keep heading south. I figured that the nearer I got to Greece and the Mediterranean, the warmer it would get. The milder climate would be good for both of us, but especially for Nala, if she had a weak chest as the vet in Montenegro had suggested. I was also determined that she cross the next border legally so, along with mapping out a route that Bogdan recommended for the journey through Albania, I was already working out where and when to get Nala her rabies vaccination.

It felt good to be organised. My mind was clearer than it had been for a long time. I had a blueprint, I could see the way ahead. Except, of course, I'd not accounted for one thing. What's that old expression? 'If you want to make God laugh, tell him your plans.' You'll make him laugh even louder if you tell him those plans involve a cat.

We set off mid-morning the following day. The weather had brightened up a wee bit and, as we took the main road down through Albania, the sun even made the odd appearance. Feeling the warmth on my face really lifted my spirits.

Nala also seemed perkier. Any concerns I had about her not enjoying travelling were long forgotten. She was so comfortable in her little pouch. When she felt like sleeping, she stayed curled up inside. But when she was awake, she'd sit up, her head poking through the open zip while the rest of her body stayed wrapped up warm. It was so entertaining looking down on her as she switched her gaze from side to side, fixing onto the latest object of interest while I cycled.

There was always plenty to see. Albania is a beautiful country, but it was obvious it had been through rough times in the recent past. A lot of the villages we passed through were run-down, farming communities. The roads were often riddled with potholes. It was a real challenge to avoid them, especially when there was heavy traffic around us. For a few miles I tried sticking to the smaller roads, but they were in an even worse state and I soon gave up. More than once I hit a pothole, sending a shockwave shuddering through the bike – and me. Luckily, Nala was better protected. Her padded pouch acted as a shock absorber.

Inevitably at one point I picked up a puncture and was forced to patch up my tyre in a field, watched by some menacing-looking goats. I had to shoo one of them away when it started sniffing at a scarf that was sticking out of one of my panniers. So, a part of me was grateful when we hit the capital, Tirana.

If nothing else, the roads were in a better state.

I don't enjoy big cities; I prefer the countryside and natural landscapes. Nala, on the other hand, was in her element. She was fascinated by the sights, sounds and smells that suddenly hit us. As we cycled past the city's giant, Soviet-era statues and buildings, and the colourful fruit and vegetable stalls, she edged out of her pouch and placed her paws on the handlebars. She was such a curious cat, she didn't want to miss a thing.

I'd decided to stay a night in Tirana to deal with a few things before heading south to the coastal town of Himara, about a hundred miles away on the so-called Albanian Riviera, which

Bogdan had suggested as a stop-off within striking distance of the Greek border. As well as being a beautiful spot to spend Christmas, Himara had vets capable of giving Nala her rabies shot before we left Albania around the New Year, when she'd be three months old.

One of the jobs on my list in Tirana was to stock up on some lek, the Albanian currency. The hostel I'd booked for us down in Himara took only cash and I'd been told a lot of banks would be closed over the coming Christmas holidays. Albania also seemed less reliable for using cards. I'd found a cashpoint in a bank and put my card in to start the withdrawal when I noticed a balance flash up on the screen. It was way lower than I expected, even allowing for the two hundred pounds that I'd forked out on my phone back in Bar.

I stood there shaking my head. This had to be a mistake.

I racked my brain. Had I used the card online where it might have been hacked? Or handed it to someone who could have cloned it? I couldn't think of anything. I stopped myself from panicking; I would have to talk to my bank.

I'd booked us into a cheap backstreet hotel with terrible phone reception, but I managed to find a spot where I could make calls and got through to the UK. The bank ran through my most recent payments, most of which I recognised. But then they mentioned two large payments a week or so earlier, totalling over four hundred pounds. They didn't ring a bell at all, even when they read out the name of the company. The charge had also been made to a company in Serbia, a country I'd not even visited.

'It can't have been me,' I said. 'I've never been to Serbia.'

'But it was a chip and pin payment. You presented your card and entered the correct pin number,' the bank person said.

It was then that the penny dropped. The vet in Budva: it was part of a group of vets. I remembered seeing a sign saying they were in Bosnia and Serbia as well as Montenegro. At the time, I'd been more concerned with getting Nala sorted. I hadn't converted the Montenegrin money into pounds; I'd assumed it wasn't too much. I'd made a mistake, obviously. A big one.

I ended the conversation with the bank feeling slightly silly. For a while I was angry with myself, but I soon calmed down. The situation wasn't critical yet. I still had a decent chunk of cash in my account and I lived frugally. I was happy to eat the cheapest food and to pitch my tent wherever I could. That was even part of the appeal of the trip. But even allowing for that, I knew I had to be careful and tighten my belt a little. That proved to be easier said than done.

By now my bike had racked up more than two thousand miles. The gleaming new machine I'd assembled back in Dunbar now looked weather-beaten and in need of some TLC. It was hardly a surprise. It had survived some gruelling conditions along the way, not least on the pockmarked Albanian roads of the past few days. The front brakes, in particular, were getting very unreliable. I had a suspicion the pads were wearing very thin. Searching online, I found a bike repair garage run by a team of young guys. I weaved my way through the backstreets of the city, checked in and left the bike with them, so they could give it a decent service. Provided I didn't need any new parts, they reassured me, it was going to be fairly cheap.

I took Nala for a walk in a nearby park and headed back to the garage around half an hour later. I could tell immediately it was bad news. The main guy in the garage looked sheepish. He told me that, while my bike was in good nick generally, they had been horrified by the state of my brakes. I knew they'd taken some heavy wear and tear, especially back in Switzerland, where the descents down some of the mountain roads had been incredibly hairy. Still, I was shocked when they produced the old brake pads. I knew the front ones had been wearing down but, removed from their holders, I could see they were practically non-existent. The rear ones weren't a lot better.

I felt guilty all over again that I'd taken Nala on the road with them like this, but I was also terrified when I asked the cost of fitting new pads. These were decent guys, though, and I trusted them. They said they'd give me the pads at cost price and only charge the minimum fee for the service. The whole bill came to around fifty pounds, which was a relief.

The bike stayed in the garage for another hour or so while I went for a coffee nearby. To be fair to them, the guys did a great job, even – to my embarrassment – giving the bike its first wash since I'd left Dunbar.

I was still smarting from forking out more cash when we got back to the hotel. I decided to get a simple takeaway and eat it in the room. I'd save some cash by having a quiet night playing with Nala and catching up on my Instagram.

As the evening wore on, I was joined in the dormitory by three other guys. Two Englishmen and, to my surprise, Bogdan, who had caught a bus from Shkodra that day. He and Nala had previously got on like a house on fire and they immediately started playing with each other. She was soon bouncing around, diving at anything she could get her young claws into. I joined in, which made her even more excited. As I pretended to chase her, she jumped off the top of a bunk onto a curtain that was hanging by the side of the bed.

She obviously planned to latch her claws into the fabric and swing on it. Unfortunately for her, it didn't work out that way. As she tried to grab the curtain, one of her claws got caught. It wasn't enough to hold her in position, so she was flipped sideways and sent flying through the air. Cats, of course, are famous for having a self-righting mechanism that ensures they always land on their feet. Nala clearly hadn't developed hers yet, because she landed on her head. For a moment there was a stunned silence in the dormitory. Poor Bogdan looked as white as a sheet.

I jumped down and dropped to my knees next to her.

Nala lay there, lifeless for what felt like forever, but was probably nearer to five seconds. It was long enough for all kinds of thoughts to flash through my head. For a split second I even thought she was dead. Before that had properly sunk in, however, she picked herself up, shook herself and limped rather gingerly into my arms. Within ten minutes she'd fully recovered, much to the relief of Bogdan, in particular.

If cats really do have nine lives, then I suspected she had now used the first of them. She certainly learned a lesson that she never forgot. I've never seen her attempt anything like that since.

Nala and I wished Bogdan goodbye the following morning and hit the road south to Himara. It was a picturesque route. The road took me through sweeping mountain gorges and past crumbling Roman ruins. I also saw dozens of old military bunkers from the Communist era. I'd been doubtful when Bogdan told me that there were three quarters of a million of them dotted across Albania. I now realised he hadn't been kidding.

The first day's cycling was fun, but on the second day the trip turned into a killer. The road was uphill virtually all the way and the gradient was so steep in parts that I had to push the bike. At one spot near a small hill village, I slowed down so much I was being regularly overtaken by elderly locals riding their donkeys. One white-haired old guy gave me a big toothless grin and a thumbs up as he and his mule went *clip-clop* past me. It was as if he was saying: *Don't give up, son, you're nearly there.* At least the weather stayed fair for me.

I rolled into the pretty coastal town of Himara around ten o'clock at night, several hours later than I'd intended. The hostel I'd booked was empty apart from a long-haired guy who seemed to be a guest-cum-caretaker. He introduced himself as Maik.

The hostel was a simple place, in a converted old building up a narrow tarmac road on a hill above the bay. But it was really comfortable with a large living area, some open courtyards with hammocks, and a spacious bedroom with three double bunks. I unloaded and put Nala on the bed, where she fell asleep within seconds. I left her to it, keeping the door ajar in case she got frightened. There was no one here but me and Maik, so it felt safe.

Maik was an interesting character. He was German, a couple of years younger than me, and said he was a traveller and DJ. We got along well and spent the first night chatting into the small hours.

He explained that the owners of the hostel, a couple, were away in Corfu for the holidays. They'd entrusted Maik with keeping an eye on the place. He was staying there for free.

'Maybe you can do the same?' he said. 'If you do a few jobs.'

I was happy to do anything that saved me money. A cheap Christmas was exactly what I needed right now.

'No problem,' I said, and he agreed to call the owners in the morning.

He was as good as his word. The next morning, Christmas Eve, he came to me beaming, his thumbs up.

'Great. What jobs do I need to do?'

'Just chop wood, squeeze oranges for the orange juice for breakfast. And keep an eye on their dogs.'

'Perfect,' I said.

I was over the moon. It felt like Christmas had come early and I was in a festive mood for the rest of the day.

Albania is part Muslim, but its former Communist government was anti-religious. So, when I took Nala for a stroll that afternoon, I wasn't surprised to discover that Christmas was a low-key affair here in Himara. There were lights and Christmas trees in the windows of a few buildings, and the shop windows were stacked with seasonal Panettone, but it bore little resemblance to the crazy, commercial holiday season back in the UK. I found it a refreshing change.

As we walked the streets, Nala was a people magnet. Several locals came up asking to stroke her and a group of teenagers took a selfie with us. One elderly lady in a headscarf spent about five minutes gazing adoringly at her as we sat on a wall, taking in the scenery. She kept muttering to herself, almost as if praying. I had no idea what she was saying, of course, but I understood one thing and it was amazing. It was almost as if Nala had a superpower, she had the ability to put a smile on faces regardless of religion, age, or culture.

That evening I left Nala snoozing safe and sound in our room with plenty of food and water while Maik and I visited a local bar for a couple of hours. The locals were very welcoming and offered us glasses of the local spirit, a fruit brandy called *rakia*. The first glass tasted like paint stripper, but I soon got used to it. Some local musicians played and the atmosphere was lovely, but quite low key. I was back in the hostel with Nala by ten o'clock.

Christmas Day was equally quiet. I began the day doing my chores, chopping wood, picking oranges from a tree in the yard, and feeding the four dogs that lived in a courtyard at the side of the hostel. One of them, an Alsatian, had a small litter of puppies and was massively protective of her brood. She growled and glared at me as I filled up her bowl.

At lunchtime I spoke to my mum, dad, gran and sister back in Scotland. We'd always been close and enjoyed our Christmases together. I could tell they were missing me as much as I was missing them. It was my first Christmas away from home. But they were happy that I had found a good base for the holiday and delighted that things had begun to take a positive turn on my trip. My dad encouraged me to press on. 'You'll only do this once in your life, son,' he said. 'So make the most of it.' It gave me a real boost.

Back home, I knew they'd be having the full traditional Christmas dinner, but I contented myself with pasta and vegetables from the supplies in the kitchen. It did me fine.

That night I watched a movie on my phone and started making fresh plans. With a fair wind, I'd get the rabies jab done around the New Year and be across the border into Greece the first week of January. Things seemed to be falling into shape again – or so I hoped.

Boxing Day was the sunniest day in weeks. The Mediterranean below us looked a tantalising deep blue. I decided to take Nala to the beach. She had great fun chasing her mouse around, much to the amusement of the locals, many of whom had come out with their families to enjoy the fine weather. A few gathered around taking photos and stroking Nala. I didn't stop them, since she would let them know if she didn't approve. As before, she seemed to enjoy the attention.

It was back at the hostel that evening, after I'd posted a photo of her playing on the beach on Instagram, when I noticed Nala breathing unevenly again. It was the familiar wheeze, but this time there was a faint little cough as well.

I felt awful. I thought the problem had gone away, that her lungs were growing stronger with age. I now regretted exposing her to that sea air.

Maik had mentioned that a vet was coming to check on the puppies during the holiday. He called him and asked if he'd look at Nala as well. The vet said he couldn't come for a day or so, but told me to keep Nala indoors and warm until he got there. I didn't need telling that; I had no intention of letting her run around outdoors, especially as the weather was forecast to turn cold again.

It was the day before New Year's Eve when the vet turned up. He was a cheery guy in a baggy, ill-fitting suit. He placed a stethoscope on Nala's chest and gave her a general check-up, ignoring me throughout while doing that classic doctor thing of shaking his head and quietly saying 'hmm' to himself. It was freaking me out, but I bit my tongue. After a while he started rooting around in his bag, and he produced some medication and a syringe.

'Cat has chest infection,' he told me in broken English, holding up the syringe. 'She need antibiotic. One now, then three weeks.'

I nodded. Poor Nala was going to look like a pin cushion, at this rate. But there was no way I couldn't have her treated for this. It had been an issue since I'd first found her and needed to be sorted out.

Nala flinched a little at the injection, but that was all. When the vet then asked me for a payment, I braced myself. Would this be the bill that cleared me out? To my amazement, when I did the calculation it was only twenty pounds or so. I happily handed him the cash.

Maik explained to him that Nala would need the rabies vaccination in January.

'If she better,' the vet told me, wagging his finger.

Afterwards I sat with Nala for a while, mulling on what had happened. My brilliant plan had been ripped up again, but this time I didn't care. I was not going to repeat the mistakes of the past. I was going to follow the vet's advice to the letter. I would put my cycling on hold and stay here in Albania until Nala was better. If necessary, we'd stay through the winter. I was going to do the right thing. For Nala, and for myself.

For the rest of that day and the following one, New Year's Eve, I kept Nala indoors. She didn't protest; cats act on their instincts and, deep down, I think she knew she had to restore her energy. To heal. Her cough had already eased off a little by the morning, but I wasn't going to rush her. While she rested, I got on with my chores.

New Year's Eve in Himara was nothing like the wild Hogmanays I was used to in Scotland. The streets around the harbour filled up with families at midnight as the bells rang, but within a few minutes they were empty again. There was no all-night partying, as far as I could see anyway. I felt a little homesick, but Nala kept me distracted. I stayed in the hostel with her and made sure she was wrapped up warm and wasn't frightened by the fireworks – not that they were scary. They lasted only a few minutes as well.

Albania was two hours ahead of the UK, so I stayed up to wait for the turning of the year in Scotland, chatting with friends and family online. My Instagram page had been filled with messages too, not only from Scotland, but also from dozens of followers all over the world, most of whom had found me and Nala in the past week.

One of them was a well-known animal website based in New York, The Dodo, who were thinking about doing a story on us.

We agreed to talk more in the New Year. I wasn't convinced they were serious; I didn't think I was *that* interesting. But it did make me wonder whether I was on to something – if I was hitting a nerve.

I wasn't sure what to do about it, though. I'd heard of people making careers as Instagram 'influencers', but had never considered that as a real option for me. Until now it had been rewarding enough to know that Nala and I were brightening up people's days. But that night for the first time I began to wonder whether it might be something I could develop more, even turn into a 'job' of some kind.

The more I thought about it, the more excited I became. Maybe I could start to do some good and somehow raise awareness of issues that I cared about, like animal welfare and environmentalism. Then I would feel as if I'd really achieved something.

Perhaps that should be my New Year's resolution, I thought to myself. *Do some good, Dean. Do some good.*

7

Noah's Ark

The two-week wait for the vet to return seemed endless. To make matters worse, the weather had turned even more wintry. I stepped out one morning during the first week of January to find a sprinkling of snow on the mountains in the distance. I was used to seeing a coating of white on the Lammermuir Hills south-west of Dunbar at this time of year, but less than an hour's drive from Greece, it seemed surreal. It kept me busy, anyway, chopping extra wood to keep the hostel warm.

Nala was happy curled up in front of the burning log fire, but I've never been one to sit still. It isn't in my nature. My mum always jokes I've got ADHD. So, despite the chilly weather, I tried to live up to the promise I'd made myself at New Year.

I began by tackling something that I'd noticed not only here in Albania, but back in Montenegro, and further up into Europe in Croatia and even Italy: the state of the coastline. It was impossible to walk along a stretch of sand without finding screeds of plastic waste strewn around. Some of it had been washed up, but even more was normal, everyday rubbish. Pop bottles, packaging, plastic bags. It made me really angry.

I'd watched a lot of documentaries on the subject and knew it was killing animals around the world. Turtles off the coast of South America were strangling themselves in discarded fishing nets. Fish and birdlife in Europe and elsewhere were choking on bits of plastic they had mistaken for food. I'd seen it close at hand, too. On several occasions, I had to stop Nala from playing with bits of shredded or broken plastic for fear she'd cut herself on the sharp, serrated edges.

A couple of days into the New Year, I set out to clear up a small beach about a mile down the coast from Himara. In twenty minutes I filled two large bin bags with a mass of plastic bottles. It then took me the rest of the morning to clear up the rest of the garbage that had accumulated there, from food wrappers and plastic bags to bits of clothing and discarded technology. Someone had even left a laptop keyboard in some rocks. It seemed as if it had been there for a while; it was caked in green algae. What genius had thought that was a good idea? I suspect the clue lay in the word *thought*. He or she simply hadn't.

I reckoned if people realised that even remote beaches like this were littered with plastic, then the scale of the problem might hit home. So back at the hostel I posted a photo of me cleaning the beach on Instagram, adding a few words to describe it. I felt a bit self-conscious about being too 'preachy'. I was no David Attenborough or Greta Thunberg, and I'd never seen myself as an activist. But I couldn't resist having a little rant about the need for us to leave beaches as we found them and, beyond that, to ditch single-use plastic for eco-friendly products.

With more than two and a half thousand eyes now on me, many of them strangers at the other end of the world, I was a little nervous about posting it. But there were soon some positive responses.

See, I told myself. *You shouldn't be afraid of voicing opinions. You're as entitled to have them as the next person.*

Over the following days, I explored the coast a little further in a kayak I borrowed from the hostel, finding other coves that needed clearing. I even went snorkelling in the ice-cold waters to see the extent of the pollution, although my main achievement there was stinging myself on a sea urchin. I had to pick the little spines out of my hand afterwards. It wasn't much fun: my hand was in pain, itching like crazy for ages.

Posting about environmental issues was all well and good, but I wasn't going to kid myself. I knew why these new people were following us on Instagram and I made sure to keep them updated on Nala and her health. She'd made real progress in the past week and hadn't lost any of her energy when it came to playtime. She raced around the hostel like a loon and only avoided the courtyard if the Alsatian was around. The dog had made its feelings towards her clear the first time they met, growling and barking at her. Nala was brave, but she wasn't stupid. She'd given the animal a wide berth ever since.

As January got under way, I felt she was on the road to recovery. She was certainly relaxed, so much so that she slept through the small earthquake that hit Himara in the first week of New Year.

I was sitting outside with Maik when it struck. The first clue came when the dogs started barking and making strange howling noises. Moments later, there was a weird rippling along the walls of the hostel, as if the building had suddenly turned to jelly. Outside in the street, house and car alarms started to go off. There were a few shouts and screams as well. It lasted for only a few seconds, but it was long enough to spook me. I'd never experienced an earthquake before. Maik told me later they were fairly common, especially further north in Albania, back up towards Tirana.

Nala, by contrast, was completely unmoved. I ran inside to check on her during the tremor and found her snoozing on a favourite sofa.

So, I was cautiously optimistic when the vet came back to check up on her and the puppies in the third week of the month. He'd remembered my request and had brought the second dose of antibiotics for Nala's chest, and the rabies injection too. I felt awful watching her have two injections, but it was for the best. I hoped there would be no need for more jabs anytime soon.

The vet seemed happy with her in general. He listened to her chest with his stethoscope and gave her the thumbs up. As the first vet had predicted, he thought that she was growing out of her wheeziness. Her lungs were getting stronger.

'I want to cycle to Greece with her. Is it okay?' I asked him.

He just shrugged.

'Why not?'

It wasn't the enthusiastic green light I'd been hoping for, but it was good enough.

I'd been in Himara for almost a month. It had begun to feel like a home away from home, too much so probably. I had the feeling I'd got too comfortable, and with the owners back soon I wouldn't be needed to look after the place anymore. So when I stepped outside two days after the vet's visit to discover the weather had turned much milder, I started packing up my gear and getting the bike ready. It was time to move on.

Packing up wasn't so straightforward these days. Getting all the paraphernalia I'd acquired not only for me, but also Nala, was time-consuming. There was so much to remember, I felt certain I'd forget something. By the time I'd swapped contact details and said goodbye to Maik, it was getting close to midday.

I left Himara with a slightly heavy heart, but also excited at the prospect of getting my journey started again. I couldn't wait to get to Greece. It had been high on my list of places I wanted to explore and felt like the gateway out of Europe, too. I hoped to head from there into Turkey, and I might even be in Thailand by summer.

I'd travelled for about an hour when Nala clambered onto the handlebars, which I'd worked out was her sign for letting me know she needed a toilet stop. I decided to kill two birds with one stone and gave her some lunch. It might send her to sleep and allow me to make better progress to the border.

I'd made myself a packed lunch back at the hostel too, but as I fished around in my bags, I realised that with a million and one things to remember, I'd forgotten it. I could still see the silver-foil wrapped package sitting on the worktop back in the kitchen. For a moment I thought about turning back, but quickly talked myself out of it. I was moving well and could already see the outline of the island of Corfu off the coast. Greece was drawing near. I couldn't turn round now, I needed to press on, but I'd have to do so on an empty stomach. I was bound to find something to eat on the road.

Sure enough, a few miles later I passed an orange grove. Fruit. That might be just what the doctor ordered. I pulled over and slid down a slope into the orange grove and reached up into one of the trees. The skins were a little leathery, but the oranges looked ripe enough to me, so I took one and peeled it. I took one bite and spat it out. It was disgusting. It was so bitter; it was nowhere near ripe.

I was still swilling my mouth out with water when I spotted something moving in a gully below me. All I could see was a ridge of brown and black markings. For a moment I was excited and wondered if it was a snake or a lizard. But when I moved a little closer, I realised it was neither.

'You're kidding me,' I said to myself.

It was a puppy.

It was as if I'd been transported back into the mountains of Bosnia. I couldn't believe it was happening again. What on earth was it doing here? There were no farms or buildings visible for miles. We were in the middle of nowhere. The puppy was really young, only a few weeks old, even younger than Nala when I'd found her. It was terribly skinny, and it was twitching and shaking, perhaps from a fever or hunger, and maybe fear as well. It must have been in real pain, because it yelped the moment I touched it.

I picked up the puppy. It weighed next to nothing. It was

breathing heavily and its coat was in a terrible condition. I suspected it had fleas or mange. Nala was playing nearby, but came scampering back the instant I appeared. The expression on her face probably mirrored the one I'd displayed minutes earlier.

What the hell?

This time I didn't hesitate and I scooped Nala up and put her in her pouch at the front. I then put the puppy in the carrier at the back. I knew I'd have to disinfect it later; I couldn't let Nala in there until then. She might catch something awful, and she'd only just been given a clean bill of health.

I took a moment to look at my map. I didn't want to turn around, I needed to keep moving along the coast towards the Greek border. The nearest town of any size was a place called Saranda. I looked online and saw the local vet wasn't open that day, but had a clinic tomorrow. It was a repeat of a few weeks earlier; I'd have to wait overnight. I could already feel a quiet dread building.

'If I take this to the vet, I'm going to be broke,' I said to myself.

But I wasn't going to abandon the puppy. I couldn't.

That night we camped in an old, abandoned garage a mile or so outside Saranda. I gave the puppy some water then tried to offer it a little food, but it didn't seem interested. It only wanted to rest inside Nala's carrier.

Nala was fascinated by her new travelling companion and kept trying to sneak up on the puppy, sniffing the air as she did so. It was as if she could smell its sickness. I made sure she didn't get too close, though. I suspected the poor dog was seriously infested with something. In the fading light I thought I could see fleas jumping out of its coat. It was heartbreaking.

I slept badly that night, my mind whirring away. For a while all I could feel was pure anger. Whoever had left the puppy out in the open fields must have known it was very ill. They had left it for dead. How heartless would you have to be to do such a thing? But slowly my mind turned to more practical questions. What was

I going to do with this latest orphan? Nala had been frail but in reasonable health when I picked her up. This pup needed urgent medical attention. It could be hospitalised for weeks.

I'd been on the road for only a few hours after nearly a month holed up in Himara. Was I going to stop for another month while this poor creature recovered? And if I did, was it then going to join me and Nala on our journey? What was I, a travelling menagerie? A modern version of Noah's Ark? No, it was a non-starter, totally crazy. But so too was abandoning it. I'd have to come up with another solution.

I got through to a vet in Saranda the following morning. He spoke English, which was helpful. Even better, he said he took in stray and sickly animals. He'd take a look at the puppy and see what he could do. He asked me to call him when I arrived in town.

My biggest concern was money. The previous night I'd mentioned this to a friend back in Scotland, who suggested starting a crowdfunding page so that people could help pay for any vet bills. I had no experience of these things, but she set it up in no time. I'd posted about it on my Instagram the night before and by the morning people were already donating money. They weren't huge sums – ten pounds here, twenty there – but it was soon mounting up. As I set off to meet the vet, it had swollen to several hundred pounds. I felt much less worried. One way or another, the puppy's medical bills would be covered.

When I rang the vet, he asked me to meet him in a square in the middle of town, which seemed odd at first. Why wasn't I meeting him at his clinic? I needn't have fretted.

He introduced himself as Sheme. He was a gentle, easy-going guy, and spoke good English; he put me at ease straight away.

Sheme took a look at the puppy and shook his head slowly.

'I don't know how people can do this to an animal,' he said, the anger plain to see on his face. He was looking more closely at the puppy when Nala appeared. She'd been skipping around the busy square, exploring.

'Where did you find this one?' Sheme asked, stroking her.

'In the mountains of Bosnia,' I said. 'Dumped by the side of the road as well.'

He smiled and shook his head.

'You and I are the same. I have four dogs at home at the moment, but I could have forty more. I want to save every stray that I pass but I know I can't.'

I nodded.

'Aye, I know that feeling.'

I could tell from the way Sheme was looking at the dog that he was concerned. He really was in terrible condition.

'It's a male, I think. No more than three or four weeks old. I will need to take him to the surgery. I can then take him home to recover.'

'Give him whatever treatment he needs,' I said. 'Money isn't a problem.'

He looked surprised. I guess I didn't look like the wealthiest guy in the world. 'It's okay,' he said. 'But if you can help find him a home, that would be good.'

He looked at his watch, as if remembering he needed to be somewhere else.

'I have to go,' he said. 'Leave him with me.'

I must have looked a little worried.

'Don't worry, he's safe with me,' he told me, cradling the puppy in his arm and stroking his head. 'I have your number on my phone. I will keep you posted on how he is doing.'

I gave the little puppy a final stroke.

'Good luck, little buddy,' I said, as they headed off.

Watching him go, I felt quietly content. I'd got the puppy to the right place. Sheme seemed like a safe pair of hands.

The weather was turning bad again. Scary-looking banks of grey clouds were building up ahead of us. The temperature had dropped a couple of degrees. So I found a sheltered spot on the coast and pitched the tent for the night. I'd strike on for Greece and Athens as soon as the weather allowed. Sheme was true to his word and sent me a note early in the evening. He'd been able to give the pup a special bath that had successfully killed off the mites that had caused the mange. He'd also given him a blow dry, some antibiotics to increase his temperature, and something to eat. The pup had then fallen asleep.

It put my mind at rest for the night. But I was still bothered by what was going to happen to him once he'd regained his strength and recovered.

That night I posted an update on the pup's progress on Instagram. I'd already come up with a name for him. Cycling along after leaving Sheme, I'd passed a rather random sign on the side of a building. It simply read Balou. I didn't know if it was a local word or surname, but it reminded me of the bear in *The Jungle Book*. It seemed to fit the puppy. In any case, it certainly went down well with my followers. Within a couple of hours, I'd had a handful of offers from people who were willing to adopt him, one of whom, a lady back in London, sounded perfect. Even better, a couple of people had asked for Sheme's details in case he had other dogs to adopt.

Their kindness restored my faith in human nature a little. It also made me feel much better about leaving him behind. Sheme had been right. If I picked up every waif and stray I found on my travels, it would take me about a hundred years to get around the world. And I'd need a giant truck to transport us all. It simply wasn't possible; I was only one person.

What I could do, though, was raise awareness.

Abandoned, abused and stray dogs were a problem all over the world. As I'd already discovered, there were a lot of decent people out there ready to offer them homes. If I could put the two together, then maybe we could make a difference. We already were, in fact.

It was a turning point; I felt another one of those lightbulbs flick on. It was the same with the plastic on the beaches. I couldn't clean up every shoreline on the planet. But I might be able to make others think about the issue, to stop littering in the first place.

As I lay there in the tent with Nala, listening to the wind howling, I felt excited. For too many years I'd been stuck in jobs where I longed for the end of the working day. Where I worked only for the wage packet at the end of the week. Now I was doing something that I looked forward to every day, that I didn't even consider to be work.

Yes, this new 'job' was going to be a little subtler and more complicated than I'd imagined. And yes, I'd have to learn the ropes, find ways to do it well. But I'd discovered something I believed in and was worth doing. I hadn't been able to say that often in my life. If at all.

8
Nala's World

One of the things I love about cycling is the space it gives you to think. Being alone on the open road, away from your phone and from other people, really helps you to clear your head. It gives you time to work things out and to deal with issues – good and bad, big and small. As I rode towards the Greek border the next morning, Nala sitting in the pouch beneath me, her ears pricked, her head darting from left to right, I found myself thinking about her. And the impact she'd made.

It was hard to believe, but in the space of a few weeks she'd shaken my life up in so many unexpected ways. For a start, my days now have shape. A routine. Today, as usual, she woke me at sun-up, licking and rubbing up against my face, meowing and demanding her breakfast. As she went outside to perform her usual rituals – sniffing and marking her 'territory' before disappearing to do her toilet – I hauled myself out of bed, spooned some breakfast into her bowl, brushed my teeth and started planning my day.

Before Nala, I got up when I wanted, or had a lie-in when I felt like it.

Those days were well and truly gone. Now I had to be on call. On Nala duty.

From dawn 'til dusk.

To do that duty properly, I've had to study a new language; to learn how to speak Nala. I picked up fairly quickly on the basic meows. The ones she uses to tell me she is hungry, tired or needs the toilet. Some of her body language was trickier, though.

For instance, at first I had no clue why, every now and again when we were cycling, she'd lean into me and lick my lips. The first time I was a little freaked out. What the hell was she up to? It felt weird. It was the third or fourth time that the penny dropped. I'd taken a huge slug of water and some of it had dribbled down my chin; she'd licked me there before doing the same again to my lips.

'Ah, I get it, you're thirsty too,' I said.

I thought I'd been keeping her well enough hydrated, but since then, she has only to stick her tongue out for me to get the message.

Aside from purring and the odd, plaintive meow when she was happy, Nala didn't have much in the way of small talk. She mainly 'spoke' when she wanted something. And she always told it to me straight, making sure I got her message loud and clear – like when she thought it was time for us to get on the bike, for instance.

My theory that Nala would simply go along with whatever I wanted, whenever I wanted, went quickly out of the window. If she was happy to see me getting the bike ready to leave, she'd jump up into the pouch and sit there, waiting for the off. Sometimes she'd even do it when I wasn't planning on cycling. She'd plonk herself in there and stare at me, as if nudging me to hit the road.

But if she wasn't ready to leave, she'd let me know in no uncertain terms. That morning was a good example. As I got ready to head off, she simply disappeared.

We'd stayed the previous night on a headland overlooking the sea, next to a little glade of pine trees where she spent much of the evening playing. I headed straight there and, sure enough, saw that she'd squeezed herself in the cleft of some branches. She didn't realise I could see her, because every now and again her head would pop up. It was hilarious. She was trying to hide.

I lured her down in the best way I know – with some of the chewy strip treats I found for her over Christmas. It took a few minutes, but in the end she came scampering over. I've learned in those moments to act fast and decisively. Before she knew it, I clipped her harness to the handlebars, popped her into the pouch, and set off down the road.

'Plenty more trees where we're headed, Nala,' I laughed, her loud meows of protest hanging on the sea breeze.

Her impact on my life was amazing when I thought about it. She'd not only changed my world, she'd changed the world around me, too. Nala drew in admirers wherever we went. Strangers would come up to her as if I wasn't there sometimes. I wasn't offended or upset; it was the reverse, in fact.

When I'd been with Ricky, we'd rarely get approached when we stopped at a roadside cafe or in a village square. We were two big Scottish guys, and probably not only looked intimidating, but also like we weren't interested in meeting strangers. So people gave us a wide berth.

I'd not thought about it at the time, but it was a shame. Part of the point of travelling the world was to meet people and learn what made them tick. How they were different, or the same as me. Cycling with Ricky, I'd missed out on that experience. Yes, we met some people, but in general they were distant, closed off. With Nala, it was completely different. She'd opened the world up to me.

All in all, I thought as I cycled along, she'd taken over my life. Become my number one obsession. That was, perhaps, the biggest change. Wherever we were now, I found myself asking a constant series of questions. Where's Nala? Is she happy? Is it time for her next meal? Is she warm enough? Where is she going to sleep tonight? It was almost like having a child. She'd become my priority, the centre of my world. I wasn't even sure if I was that important, come to think of it. Maybe I was the centre of hers?

I looked down at her, sitting proudly in the pouch, like the commander of our little ship, surveying the world from the upper deck while I pedalled away in the engine room, Scotty to her Captain Kirk.

'Aye, that's it,' I laughed to myself. 'It's Nala's world now, I'm just living in it.'

As we approached the border patrol, I felt strangely nervous. I had all the documents I could possibly need. Yet my head was playing tricks on me again. Was the paperwork going to be correct? Would they find some little detail that would keep Nala out?

I pulled up at the window and handed over the two passports. The guard was a big, moustachioed guy in a slightly scruffy uniform. He seemed more interested in the game of football that was on a little TV in the corner of his booth. He looked at me and then at Nala, who was sitting upright in the pouch staring at him, her head tilted to one side. It was as if she'd decided to turn the cuteness dial up to maximum.

Flicking through our paperwork, he looked baffled. I suspect he'd never seen a pet passport before. After no more than thirty seconds, he shook his head and stamped my passport, a thin smile breaking across his face as he did so. He shoved my documents back at me and waved me through, winking at Nala as he did so.

I couldn't help chuckling to myself. This had been our first official border crossing together and that was it? All that worry. All that effort. All that money to ensure I had the correct paperwork for Nala. All I'd needed in reality was to show her adorable, irresistible face. It was a passport in its own right.

For a long time during my journey to southern Europe, I'd naively assumed that Greece was a land of constant sunshine. I think I half-expected people to be walking around in T-shirts in January. It didn't take long to shatter my illusions. My plan was to cycle across country and cut down towards Athens. As we got under way, the weather was bitingly cold. There was a sharp wind blowing in from the north, where I could see fresh snow on mountaintops.

Winter was far from over.

It wasn't long before Nala was attracting attention again, this time from some unexpected quarters too.

She'd had a good sleep during the middle of the day, so was sitting upright in the pouch, surveying the countryside. As we cycled through a succession of little farming villages, people would point and smile. When we went past a small schoolyard, the kids in the playground waved and shouted.

I gave her neck a little ruffle.

'It's like travelling with bloody royalty,' I told her.

But then, ten or fifteen miles into Greece, I noticed a police car looming up behind us. I waved it past, but it stuck there, as if it was following us.

'Uh, oh,' I said to myself. 'This could be trouble.'

As I feared, a mile or so further down the road, there was a flash of headlights. When I looked round, I saw there was a single officer in the car. He was waving at me to pull over.

I slid off the road and came to a stop next to a small church.

It was like a scene from a movie. I watched as a short, middle-aged guy in a dusty, dark blue uniform got out and walked slowly towards us. I noticed straight away that he had a gun belt strapped to his hip. I didn't think I'd broken any laws, but I started going through my document pouch just in case.

The policeman walked straight past me to the front of the bike.

'Your cat is beautiful,' he said, bending over Nala. 'What is her name?'

'Nala.'

'Hello, Nala,' he said, gently stroking the back of her neck.

I had my passport in my hand and offered it to him, but he waved it away.

'Where are you going, my friend?' he asked.

'Athens. Eventually,' I replied. 'Just going to take a few days to cycle down there.'

He pointed to the mountains and the steely grey skies above them.

'Very bad weather coming. If I were you, I would camp in the next few miles.'

'Okay,' I said. 'I'll do that.'

'Don't want your cat to catch a cold.'

He leaned down and stroked Nala one last time, giving her little pretend kisses as he did so.

'Safe journey, Nala,' he said, before nodding at me and heading back to the car. A few moments later, he'd driven off. Just like that.

I shook my head in disbelief. Had this policeman seriously pulled me over to warn me it was going to rain? Or had he really wanted to say hello to Nala? I couldn't make up my mind. One sounded as crazy as the other. As it happened, the policeman's advice was spot on. By the time I rolled into the next decent-sized town, the weather had broken. I got the tent up as fast as I could and curled up in the warm with Nala. We were soon inside listening to the drumming of the rain. I contented myself with getting on with some work, finalising the plans I'd begun making for when we got to Athens.

With money tight, I'd visited a couch-surfing website that Ricky and I used a couple of times when we first started out. Some free accommodation would ease my worries. I'd already got an invite to stay up in the north with some friends of an aunt back in Scotland, at a place called Neos Skopos, on the way to Thessaloniki. But I'd also put up a request to stay somewhere in Athens for a couple of nights. Since they call Edinburgh the 'Athens of the North', I thought I should see the original.

I quickly had an offer from a family who said they'd be delighted to host us. I assumed they'd done it purely out of kindness, but I should have guessed they had an ulterior motive. The mum admitted their daughter was a massive cat lover and had spotted in my profile on the website that I travelled with one.

'She can't wait to meet Nala,' she said in an email.

Unfortunately, Nala hadn't helped me with my other quest – to find some work. I'd decided that I needed to earn more money. Nala was an extra expense and I didn't want to be caught out if she became ill on the road. She'd be six months old in early April and would need to see a vet again for new jabs and maybe even to be spayed. I was determined to keep the GoFundMe money for looking after Balou, who was still being treated by Sheme back in Albania. So I replied to some ads with jobs for a kayak guide.

I'd been in and out of kayaks since I was a teenager. So I sent off applications to about a dozen different companies. A couple had replied to tell me they were fully staffed for the coming season. Most of the rest hadn't even bothered to respond. But I was determined to stick at it. Maybe there was one out there who wanted a mouse-catcher as well?

Our progress during the next week was governed by the Jekyll-and-Hyde weather. It could flip from gorgeous sunshine to thunderstorms and back again in moments. When the sun did shine, there was no doubt about it, northern Greece was paradise. We travelled down the north-western region of the country spending our evenings camped on the edge of deserted coves, watching seabirds hovering over the waters and listening to the waves crashing in. But when the rains came, it was like Scotland. Grim and grey and soggy.

I didn't mind. It was great to be on the road, moving from place to place. It was the kind of cycling I loved, the kind of back-to-nature living I relished – though I hadn't planned on having a playful cat to share it with me.

Nala was growing fast. She was still a wiry wee thing, but the days when she'd fit in the palm of my hand were long gone. She'd also grown more and more confident about leaving me.

Provided she felt safe, she'd wander off quite a distance. It was funny watching her when we arrived somewhere new. She'd start by checking the place out, like some police forensics officer, sniffing the territory as cats do and picking up on the smell of other creatures; rubbing up against things to lay down her own scent. I know now that cats can tell an incredible amount from smells. For them, it's like reading a guidebook or looking at a map. As soon as Nala was happy with her surroundings, she would get on with the job of enjoying herself.

She was such an athletic cat, she thought nothing of jumping and climbing obstacles that I'd have thought impossible. It was another one of her superpowers, I decided. She could leap ten feet straight up from a standing start, as if she had powerful springs in her legs. She was fearless, too. I liked to think it was one of the ways in which we were similar.

Also like me, however, she had developed a habit of getting herself into trouble, and of biting off more than she could chew.

One evening that week, as I was putting up the tent in a sheltered spot overlooking another idyllic stretch of coastline, I heard her meowing loudly.

For a while I ignored it. She'd had her dinner and something to drink. She'd done her toilet. What more could she want? But it didn't go away; in fact, it only got louder and more intense. She began meowing in short, sharp bursts. There was almost an anger to them. If I'd not known differently, I'd have thought she was swearing.

I left the tent half-assembled and went off to investigate. I knew that she'd headed for a copse of trees. So I scanned the lower branches, but there was no sign of her.

The meowing started again. Louder than ever this time.

I could hear it coming from above me. I looked up and there she was, balanced on the edge of a flimsy-looking branch at least twenty feet up.

'How the hell have you got up there?'

She must have climbed up the main branches of the tree, but then jumped out onto this branch without realising the risk. She'd literally put herself out on a limb. The problem was that the lighter branch was moving around wildly in the breeze, and Nala was frightened of jumping back onto the main branch of the tree.

I did a quick recce, then clambered about a third of the way up the trunk of tree, balancing on a thick, knotty branch. I then grabbed hold of another, lighter branch, which rose up at an angle to the one where Nala was stranded. It was flexible and easy to manoeuvre, and I began moving it up and down so that it came close to Nala. It offered her an escape route, a ladder down to me.

She didn't seem too keen at first and hovered on the edge of her branch, stepping forward and then back again, as if plucking up the courage to jump.

'Come on, Nala,' I shouted a few times. 'You can do it.'

Finally she went for it, scampering down the escape ladder like some high-speed tightrope walker. She didn't even bother coming to me, she carried on all the way down to the ground, landing with a bump.

By the time I dropped myself back down to earth, she'd dusted herself off and trotted away as if nothing had happened.

'My pleasure, your highness,' I said, as she walked back towards the tent. The brief meow she let out wasn't one I recognised. I suspected she might have been swearing again.

It took us a week to cycle down to Athens. My usual dislike of big cities fell by the wayside almost as soon as I saw the Parthenon looming over us. I was bowled over by the sense of history and energy of the place. Every corner seemed to have some crumbling, ancient monument or statue on it. But it was a vibrant, modern town, too.

The family who'd agreed to put us up lived in a lovely, leafy suburb near the centre of the city. Approaching the smart-looking house, I was a bit apprehensive, especially after a few rough days on the road. For a start, I was in need of a decent shower. But I also had no idea what to expect. They seemed nice enough in the emails, but you never knew for sure who you were going to encounter when you couch-surfed like this.

Luckily, the mum and dad, Nick and Iliana, and their daughter Lydia, couldn't have been kinder or more thoughtful. They welcomed us as if we were long-lost friends. The day after we arrived, Nick even got in contact with the city's branch of Trek, the makers of my bike, and arranged a service that he insisted on paying for. I was dumbstruck that people could be so kind to a guy whom they didn't know from Adam.

Nala took to them instantly as well, which was hardly surprising given the fuss they made over her. Lydia would have played with her every minute of the day if she could. The pair of them tussled and wrestled and cuddled in front of the TV as if they'd known each other all their lives.

I loved watching them together. It gave me the odd break from Nala duty and some time to catch up on other things.

Since leaving Albania, I'd stayed in regular contact with Sheme. Balou had suffered a setback and had to be put on an IV drip for a while. But he'd finally turned the corner and was now on the road to recovery, and being prepared to go to a proper dog trainer back up in Tirana. The lady from the UK who was adopting him was also planning to visit him. It was all heading in the right direction. Although I was happy, I was frustrated that I'd only seen him when he was ill. At the rate things were going, he'd soon be on his way to London. I might never see him again after that.

Nick and Iliana had followed me on Instagram and asked about Balou over dinner one evening. I must have done a bad job of hiding my feelings, because they immediately picked up on how conflicted I felt.

'Why don't you go up and see him,' Iliana suggested.

Nick nodded and smiled.

'We can look after Nala for a day or two,' he said. 'Can't we, Lydia?'

She almost cheered with delight at the prospect.

It took me aback. I'd never even considered it.

'Really?'

'Really,' they all said.

I looked into it. There was a bus from Athens' main coach station that would take me back up through the border and directly to Saranda. I could go overnight, spend a day with Balou, then get the return bus overnight. I'd be away for around thirty-six hours in total.

I felt really bad leaving Nala behind. We'd not been separated since the mountains in Bosnia. But I knew she couldn't have been in safer hands. When I left late one afternoon a couple of days later, she was playing with Lydia and barely even noticed me sliding out the door.

The six-hour journey up was hellish. The heating on the bus had been turned up to maximum. At one point I thought I was going to melt. I finally got to Saranda in the morning and headed straight to the spot in the middle of town where I'd met Sheme previously.

He arrived after a few minutes. The lively-looking dog he had at the end of a lead bore no relation to the Balou I'd left here weeks earlier. He was two or three times the size of the dog I'd found by the road and looking really healthy; his coat was shining and he was moving around easily. Sheme left him with me for a couple of hours, so I took him for a walk around the town. It felt great to have him tugging on the leash, diving into bits of shrubbery on the side of the street like any normal dog. It was plain to see that he was going to be fine.

I didn't want to leave Nala on her own for too long, so I turned around that night, saying a happy farewell to Balou and Sheme before catching the same bus on its return journey. It felt as if I'd brought a chapter to a close. The chances were I'd not see Balou again, but who knew? What I did know was that I'd played my part in giving him a new start in life. And that felt very good indeed.

The heating on the bus was turned up to the max again on the return journey, but that was the least of my concerns this time. I'd become almost blasé about border crossings, so when the Albanian police decided to pull our bus over to one side, it was a rude awakening. I didn't know if there had been a tip off about drugs or something else illicit, but they went through the bus with a fine-tooth comb, putting everything – including the bus – through an X-ray. Not once, but twice. At one point they hauled two young men off into a small building at the side. They did not return.

All in all, we were there for a couple of hours. It gave me time to think. At our last border crossing together, I'd almost dismissed the requirement for proper paperwork. All I needed was Nala's pretty face. Who was I kidding? There was no way we'd escape an interrogation like this if we didn't have all the official require-ments. If I was to take her further around the world, I'd need to stay on top of it all the time. Every country was going to have its own rules, and I couldn't risk having Nala hauled away like those two guys.

Back in Athens, Nala shot out and jumped into my arms when I walked through the door. She was purring heavily and rubbed up against me. I held her tight in my arms for a bit, but after a while she looked at me, as if saying: *Okay, mate, I didn't miss you that much.*

She soon ran off to play with Lydia again.

As February drew to a close, I was getting despondent about my job situation. I'd sent out a new batch of applications to kayak centres all over the Aegean and Ionian islands. A couple of places had replied, one asking a few questions about my experience and certification, but I didn't have any official qualifications. I'd begun to think about coming up with a Plan B. Maybe I'd need to apply for some bar work.

I'd come back for lunch one day after a cycle around Athens with Nala, when another email popped up. It was from a guy called Haris, who ran a kayak business on the island of Santorini, in the southern Aegean. He said they might have a job for me, but only if I could start as soon as possible. He and his brother needed help setting things up for the summer and wanted to know if I could start at the beginning of April.

It was short notice and I didn't know the Greek islands that well, so asked Nick and Iliana. They both smiled.

'Santorini? It's one of the most beautiful places in Greece,' Iliana said.

'In the world,' added Nick. 'You must go there if you have the chance. You can get the ferry from here in Athens.'

That made my mind up for me. I wrote back saying I could get there by the end of March. I then began looking into ferries from Athens. Boats went from the main port Piraeus to Santorini a few times a week. Crucially, they accepted cats, provided they were in a proper carrier.

With under a month until I would sail, I decided to get back on the road to explore more of northern Greece. Even though they'd not said anything to suggest it in any way, I felt that we'd overstayed our welcome with Nick and Iliana. We'd originally said we were going to stay for two days, but had been there more than two weeks. Now I planned to check out Neos Skopos and Thessaloniki, as well as the famous hot water springs in Thermopylae, which had always been on my list of places I wanted to visit. I'd head to the springs first.

I had a suspicion that getting Nala away from Lydia might be a challenge. Sure enough, when she saw me loading up the bike, she scuttled back into the house and hid under the sofa. Lydia had to help me tease her out and then get Nala into her pouch. It took twenty minutes.

The final parting was almost as long and felt like something from a Hollywood tearjerker. The whole family were crying. I wouldn't have been surprised if an orchestra had started playing soppy music in the background. Somehow, I managed to hold it together; if I hadn't, then we'd probably still be there today.

'Don't worry, you'll see them again,' I told Nala as we cycled off. The low, rasping growl she gave me was unlike any noise I'd heard her make before. I didn't think I wanted to translate it. I was sure it wasn't a compliment.

9

A Blessing

We arrived at the hot springs of Thermopylae a couple of days into March. It wasn't hard to tell we were drawing near. The muffled roar of the waterfalls and the faint, rotten-egg smell of the sulphurous waters built steadily as I steered my bike along the twisting mountain road. The springs were adjacent to the famous pass where Leonidas and his 300 Spartans held the Persian army, two and a half thousand years ago, so the area was a major tourist attraction during the peak season. Fortunately, summer was still three months away. As I cycled past the statue of Leonidas and the official visitor centre and museum, and into the car park next to the springs, I could see only a few cars with overseas number plates from France and Sweden. Otherwise it was quiet.

The main springs are a network of natural and man-made pools and baths arranged around waterfalls and a rocky stretch of fast-flowing river that weaves its way through the surrounding woodland. With Nala on my shoulders, I set off to explore and soon found a gentler section of the river that was almost empty. Nala seemed suspicious of the slightly milky, turquoise waters, especially as there was steam rising off them. They also smelled really strong. She kept sniffing the air, as if thinking: *What is that stink?* When I stripped down to my shorts and dipped myself in, she looked on in disbelief.

Nala didn't know what she was missing. It was amazing, the waters must have been 40 degrees: it was like stepping into a warm bath and a massive, natural hot tub at the same time. I have to admit, I needed the bath after a tough week on the road.

The cycle there had taken four days or so and had been gruelling. A couple of nights after leaving Athens, I'd camped in an old hill fort overlooking a sweeping valley. It hadn't been ideal for pitching a tent, so Nala and I had slept under the stars. As we curled up in my waterproof bivvy bag, the skies above us were as clear as a bell, but at around five in the morning, I was woken by a violent thunderclap. It was as if a bomb had gone off right next to us. A storm had settled right over our heads and proceeded to dump raindrops that were so big that I could feel every one of them hitting my bag.

Thankfully, the storm was as short as it was violent and my bivvy bag kept the water out. Nala stayed tucked underneath me throughout, asleep and unfazed. But I felt a bit sorry for myself, lying there in the storm. I'd felt something similar two days later, when I decided to go off the main road and ended up cycling down a muddy track. By the time I'd hauled the bike through, I was caked in the stuff. It was only now, as I lay in the warm waters of the hot springs, that I was able to scrub off the last of the mud.

People had been coming here for thousands of years, drawn by the supposedly healthy qualities of the sulphurous mineral water that had been driven up from the earth's core. By the time I'd finished, I certainly felt a hundred per cent better, and cleaner than I'd been in a while.

There was a lot to see in the area, so I decided to camp here for the night. A short distance from the hot springs, I saw what looked like a hotel or hostel of some kind, set back in the woods. The yellowing building stood alongside the river that flowed from the springs, and it had seen better days. The grounds were untidy and overgrown.

There was an area of open, scrubby land in front of the building that looked perfect for my tent, but I couldn't work out whether it was within the hotel grounds. I didn't want to go to the bother of pitching my tent, only to get kicked out again.

A young lad was standing next to a wall at the entrance to the hotel's driveway, selling fig jam to tourists. He saw me and started waving his arms.

'It's okay. It's okay,' he said.

'You sure I can camp here?' I asked him.

'Sure. One hundred per cent. No problem.'

I'd always had a sweet tooth, so I bought a small jar of jam off him, as a thank you. He was very grateful; I sensed he'd had a quiet day.

While Nala explored the area, I started clearing the ground for the tent. I had noticed a few children playing in front of the hotel and I'd barely got the tent up when a couple of them appeared. They were young, dark-haired girls, dressed in tracksuit bottoms and hooded tops. As ever, it wasn't hard to figure out what had attracted them.

Tired out after exploring the area, Nala was tucking into some food that I'd laid out for her. I waved the girls forward and they kneeled down a few feet away from her, leaning and giggling as they chatted in a language that didn't sound like Greek.

I told them her name, which they seemed to understand.

'Nala, Nala,' they said. I'd never seen the cat turn down the offer of some attention, and she was soon playing with them.

After about ten minutes an older girl appeared, shouting and waving at the youngsters. It was clearly their dinner time. The older girl seemed friendly and acknowledged me, then she led the girls away. They kept turning and waving as they headed back, shouting, '*Ma salama*, Nala!' I assumed it meant goodbye.

I'd read that sulphurous water was safe to drink, despite its rancid smell. But I knew Nala wouldn't go near it, so I headed to the hotel to see if I could fill up my water bottle there, figuring they might filter the local water.

The closer I got, the more obvious it was that it wasn't a normal hotel. There were long lines of washing hanging from most of the balconies. People were sleeping in hammocks on some of the others. Next to one of the rooms, a group of women were cooking what appeared to be a rabbit on an open fire.

Inside, the main entrance hall was stripped bare and there was hardly any furniture, apart from a couple of old leather sofas in a corner, where some men were watching TV in what sounded like Arabic with the volume turned high.

I was about to approach them when a young guy appeared, carrying tea in small glass cups. He seemed surprised to see me.

I waved my water bottle at him and pointed at Nala, on my shoulder.

'For my cat,' I said.

It turned out he could speak a little English.

'Ah, okay. Water,' he said, relaxing a little then handing the teas to the men on the sofas. 'Wait.'

He then gestured for me to follow him.

'Come with me.'

He took me to a small kitchen area, where he poured some water from a large, plastic tank. It looked drinkable.

The kitchen was also in a bad state of repair. The walls were stained and the paint was peeling off. The kitchen equipment looked rusty, as if it hadn't been used in a while.

'So, what is this place?' I asked the guy.

'It used to be a hotel,' he said.

'Used to be? What is it now then?'

'Oh, we are refugees. This is now a refugee camp.'

I was taken aback. I'd always assumed, naively, that refugee camps were grim places with high wire fences and hundreds of tents. Rough and ready accommodation. This lodging, while on the basic side, was in an amazing location. How it felt to be confined here in these circumstances was another matter, of course.

I wasn't sure what to say, so I thanked the guy for the water and headed out. I took the long route back to my tent, walking around the grounds. There was a strange atmosphere.

On the one hand, the children here seemed to be very happy. A group was jumping off rocks into the roaring waters of the river, which ran directly alongside the building. Another group was playing football against a wall near the back of the hotel. They looked like they didn't have a care in the world. But elsewhere, groups of adults were huddled together, sitting on rocks or broken old chairs, in quiet conversation. By contrast, they looked like they carried the weight of the world on their shoulders.

I didn't want to intrude, so I headed back to the tent to eat with Nala. It was nearly dark now. It had been a long cycle up here and I soon drifted off. Nala woke me up as usual the following morning, a little later than normal. When I unzipped the tent to let her out, I did a double-take. A plastic bag had been placed next to the bike. I couldn't believe it when I looked inside. It was full of oranges, tomatoes, bread and water. A full breakfast, essentially.

'What the hell?'

It must have come from the occupants of the camp. They might not have much, but they'd still shared their food with me. I was stunned.

It was a bright morning and I sat outside the tent, feasting. The bread went really well with the fig jam I'd bought.

The chain on my bike had been behaving a little oddly on the climb up here, so I decided to spend the morning doing a mini-service. No sooner had I got the bike up-ended when the little girls from the previous evening returned, this time with a couple of new friends.

'Nala, Nala,' they kept saying excitedly to the new girls.

I let them play with her and took a photo of us all. The girls looked happy and contented; you'd never have thought they were in a refugee camp. I guessed that maybe life here was better than whatever they'd left behind back home. I dreaded to think what they'd been through to get here.

I was still fiddling with the bike when I was joined by a middle-aged man wearing jeans, a sweatshirt and a battered old baseball hat. He seemed to know the girls and spent a moment talking to them before coming over to me. He was soft-spoken and appeared to be well educated. He certainly knew decent English. He asked me if I needed any help, but I said I was okay. He then sat down, cross-legged on the grass next to me.

'Where are you from?' he asked me.

'Scotland.'

'Ah. Bagpipes,' he smiled, squeezing his arm into his armpit and pretending to blow, in what I assumed was a very poor impersonation of a Highland piper.

'That's right.'

'And where do you cycle to? Athens? Thessaloniki?'

'Around the world. With my cat over there,' I said.

He turned and looked at Nala, playing with the little girls, then smiled at me.

'In my religion, Islam, it is said the prophet Muhammad preached with a cat on his lap. When he found it asleep on the sleeve of his robe one day, he cut the sleeve off so that his cat could rest in peace,' he said, a broad smile on his face. 'So now it is said that love for cats is a sign that you are a believer.'

I nodded. It explained the reaction to Nala I'd seen in Albania and elsewhere.

A thought seemed to occur to him. 'You will go through Turkey?'

'Yes. That's the plan.'

'They will like your cat there,' he said, the smile fading now. 'But be careful. Stay to the north, do not go near the border with Syria.'

'Is that where you are from?'

He nodded his head slowly.

'It is very bad there,' he said, looking down. 'Very bad.'

In the past couple of years, I'd seen a lot on the TV news about Syrian refugees risking their lives to escape the bombs and bullets, then braving it on small boats to cross into Greece. It looked hellish. I couldn't begin to imagine what they were going through.

'Is that where everyone here is from?' I asked him.

'Not everyone. Some from Iraq. Some Kurds. We are all trapped here.'

'Trapped?'

'We cannot move. We want to go to Germany. Or Sweden. Or Scotland,' he smiled. 'But no countries between here and there are willing for us to cross their border.'

I'd heard a little about this. A lot of the Balkan countries had put up barriers to stop refugees passing through. So, having got this far they'd found themselves stuck, unable to go on to northern Europe, but also scared of going back home.

'Quite a nice place to be trapped, I guess,' I said, trying to make light of things.

He looked around him.

'It was a hotel,' he said.

'Aye, so someone told me,' I replied.

'The Greek government turned it into a centre for us. We have made it a home. We have a small library for the children,' he added, smiling at the girls playing nearby. But his face had soon turned more solemn again.

'But maybe not for much longer. They will move us soon. Maybe we have to go back to Turkey, or another camp.'

For a moment there was a silence between us. I wasn't sure what to say. What could I say?

He broke the silence after a while.

'So, will you cycle to Australia?'

'Australia? Maybe, yeah. One day.'

'I would like to go to Australia. See kangaroos,' he said, getting up on his haunches and breaking into another impersonation, this time of Skippy. For a moment he chuckled away to himself, as if pleased with his little joke.

I offered him one of the oranges that I'd been given.

'*Shukran.* Thank you,' he said, accepting it with a smile.

I peeled one open myself. It tasted a lot sweeter than the one I'd grabbed back on the roadside in Albania.

'No, it's me who should be thanking you,' I replied. 'It was someone here who gave these to me. There was no need.'

He nodded towards the little girls, then looked at me for a moment.

'Be a blessing to others and you will be blessed,' he said.

Beyond letting the girls play with Nala, I hadn't really been much of a blessing to them, or I didn't think so. But I got his meaning, and I wasn't going to argue with that. I believed it too.

He stayed there for a little while longer, slowly eating his orange, watching me working on the bike. I supposed I was a distraction, something to break the monotony. But eventually he moved off, making a little prayer gesture and nodding his head as he went.

It carried on like that all day. Every now and again, I'd pop into the tent for a lie-down out of the sun. But I'd have hardly settled down when there was another head poking through, offering water for Nala or merely curious to see the man with the tattoos and the strange Scottish accent. It was as if Nala and I were the camp's new entertainment. I didn't mind at all. I was glad to lighten the mood, especially for the adults. It was all down to Nala, of course. No one would have come over to talk to me if she hadn't been there. She really was opening my eyes to a wider world.

After lunch I played with the kids for a while, kicking a battered and deflated football around. The ball would have been thrown in the bin back home in the UK. But these kids couldn't have cared less. I then dug out the frisbee I carried with me, teaching them how to throw it to each other. Nala was in her element, chasing and trying to catch it mid-flight as the kids screamed and laughed along.

I'd really taken to them, so midway through the afternoon, I hopped on the bike with Nala and zipped back down the road to the shop I'd passed on the way up. I bought a load of chocolate and sweets, which I then shared around. They gobbled the sweets down as if it were their birthdays. It felt good to see the smiles on their faces. I had no idea how often they were given a treat like this, but it was like the Syrian man had said. Someone within the camp had been kind to me, it was the least I could do to return the favour.

The four girls stayed there until the sun began to set behind the massive mountains to the north, and their mums called them in for their evening meal. As they began to scatter and run off, I gave one of them the frisbee to keep. She didn't understand at first, but when it sank in, she gave me the biggest smile.

That night I put the photo I'd taken with Nala and the little girls on my Instagram. I didn't engage in any political chat; I didn't feel qualified to do so. I simply explained what a lovely day I'd had with the kids and the others at the centre. That was the best I could do. If it made even one person more aware of their situation, then it had been worthwhile. One more other than me, that is.

I set off early the next morning, but couldn't stop thinking about the people at the camp. Here they were, people who'd been ripped from their own homes and communities, who'd been stripped of all their possessions. But even though they had nothing, they still had something to give. It was humbling beyond belief.

'No more complaining from you – or me,' I said to Nala, as we cycled along.

My accidental encounter with the refugees played on my mind throughout the rest of the journey north. I headed next to my Aunt Helen's friends in the quiet little town of Neos Skopos. Like my Athens 'family', they were amazingly generous, even finding Nala and me a little house to stay in a few streets away. They also

Me, with the seagull
I rescued when I was a
schoolboy in Dunbar,
Scotland.

With the family. Left to right: me, my
dad Neil, mum Avril and sister Holly.

Ricky and me, preparing to hit the
road during the autumn of 2018.

Playing with Nala on a Montenegro beach soon after we met.

On the city walls in Budva, Montenegro, in December 2018.

Nala settled in for the ride to Tirana, Albania.

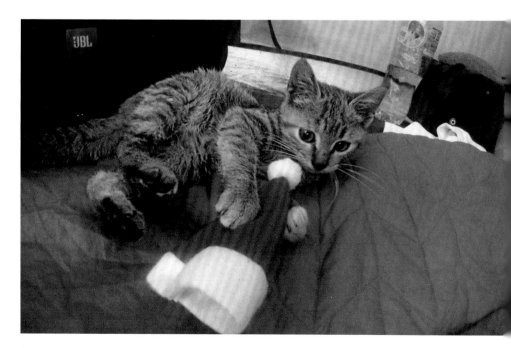

Nala's first Christmas: December 2018 in Himara, Albania.

Diving after her mouse toy on Boxing Day 2018 on Himara beach.

Saranda, Albania, after dropping Balou off with Sheme, the vet, in January 2019.

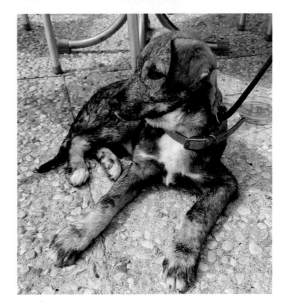

Balou. A picture of health in Tirana, a few months after being rescued.

Nala's first night in Greece, January 2019.

Taking a break in Thiva, Greece.

With Nick, Iliana and Lidya in Athens.

With Tony at the kayak base on Santorini.

Nurse Nala. Helping me recover from illness.

'It's for your own good.' Wearing a makeshift collar after a visit to the vet.

A little TLC to help Nala recover from her neutering operation.

Getting used to the water and the kayak.

Being spoiled at her local restaurant on Santorini.

Making herself at home in the kayak.

Meeting some of the dogs at the SAWA charity near Fira.

Making pottery to be
auctioned for charity.

Next stop, Asia. Catching
the ferry from Chios to Çeşme,
Turkey, in July 2019.

Settling down for the night in a disused
swimming pool on the road to Izmir.

invited us to a massive Greek Ash Wednesday feast to mark the traditional beginning of forty days of Lent. But even as I dived into the plates of humous and taramasalata and pitta bread, I couldn't shake off thoughts of the contrast with the lack of food at the camp – no matter how hard I tried telling myself that there was nothing I could do about it.

It was the same when I cycled across to Thessaloniki, Greece's second biggest city. I had a couple of nights sleeping in the tent *en route*. The rains fell on us again, but this time, if I felt even slightly sorry for myself, I repeated the same mantra.

What had those kids been through? Where had they had to sleep?

We arrived in Thessaloniki during a break in the wet weather. It gave me a chance to explore the city, a striking mix of the ancient and the modern. It was once one of the greatest cities in the Byzantine empire. With Nala on my shoulder, I visited some of the ancient monuments, including the Roman Agora and Rotunda as well as the famous Arch of Galerius, a monument to a famous military victory.

Nala, naturally, was more interested in exploring the parks and the open spaces. She'd developed a real fascination with watching birds. She'd make this odd clicking sound as she looked up at them on the branches of trees. I think she was considering making them her supper, but I didn't let her stray off the lead. The consequences didn't bear thinking about.

I had a few things I wanted to deal with during my stay in Thessaloniki – some serious, others less so.

I had my first tattoo on my leg when I was around nineteen. It was done in Newcastle and didn't mean anything, it was only a pattern. But since then I've added another ten or so, many of them to mark significant things in my life. I've had some of the lyrics of Eminem's 'Till I Collapse' inked on my torso. So I'd been thinking for a while about having a tattoo in honour of Nala. She'd become a part of me, and I wanted to acknowledge that.

I found a good parlour and got the young, female artist there to ink Nala's paw-print on my wrist. I thought it was best to put it where I could always see it. I was more than pleased with the results.

Back at the small hostel I'd booked us into, I finally took a moment to talk to the lady at The Dodo website, Christina. It felt weird talking about myself at first and I was always self-conscious of my accent; I didn't expect her to be able to decipher my Dunbar brogue. But by the end I'd relaxed more and was happy to talk through the detail and drama of my first meeting with Nala. Christina said she'd be keen to use some of my video footage in the piece they were hoping to publish.

It took me a while to sort through the old clips and get them across to her. At one point when the internet was playing up badly, I almost gave up, but I managed to send them. Christina hadn't made any guarantees about running a piece on me, so I quickly forgot about it. My reasoning was simple. Who would want to read or watch a video about a scruffy layabout from Scotland and his stray cat?

The trip back down to Athens was fun – and eventful.

I stopped off in the town of Volos, where I couch-surfed with a lady named Felicia. She was incredibly kind, even taking me for a night out on the town with her friend Yamaya and some others, which I appreciated. I didn't want to get back into the partying habit, but I still needed to let my hair down now and again.

As I made my way further south, I had a few more hiccups. Trying to cross a river, I managed to let the bike fall over on its side, soaking all my kit in the process and giving Nala a scare. I had to dry my stuff out on the riverbank, but I didn't mind. I'd chosen to make this journey – so how could I feel sorry for myself when I'd seen people whose lives had been turned upside down through no fault of their own.

My route back towards Athens took me past Thermopylae and the camp. I felt drawn to say hello, but as I approached, I saw there was a lot more activity than a week or so ago. Dozens of people were walking along the road with bags and rucksacks. There were some official-looking people around as well, a few of them in uniforms.

I biked up the road to the entrance near where I'd camped previously. As I slowed down, I heard voices shouting.

'Nala! Nala! Nala!'

Some of the kids had run out from the old hotel to say hello, including a couple of the girls we'd met before. They gathered around the bike and one or two gave Nala a little stroke.

A part of me was tempted to stay, to pitch my tent and spend some time there. I felt that I wanted to know more about them, to hear people's stories in detail.

But the children were soon being called back. Something was changing. Outside the hotel, I could see a few families gathered with their belongings in front of them. Maybe, as the Syrian man who shared my oranges had predicted, they were being relocated. I couldn't help wondering where they'd end up next; and what would happen to them when they got there. It was a disturbing thought.

I waved the children – and the camp – goodbye, hoping the very best for them.

The following day was my birthday.

I camped out at a beautiful spot overlooking the coast and spent the day chilling with Nala and chatting to my mum, dad, and sister back in Scotland. Perhaps it was the impact the camp had made on me, or it was because I'd spent a fair amount of time around other families in the past month, but I felt more homesick than usual. It was the first birthday I'd spent away from Dunbar. Talking to my family made me feel better; it was good to catch up on each other's news. They'd been worried about my money running out, so were delighted that I'd soon be heading down to Santorini for the summer. My mum had also made a cake, which the three of them proceeded to eat as I watched on my phone.

'Don't waste any. There's people who'd live off that for a month,' I said.

'God, you sound just like your father,' my mother laughed.

She was right. When me and my sister were kids, he would often preach to us about starving people around the world. Tell us how well off we were in comparison. Like most kids, I'd paid little attention, but now I understood.

It took me another four or five days to get back to Athens. I then had a couple more days before the ferry for Santorini, so I accepted an invitation to go back to stay with Iliana, Nick and Lydia. Lydia, in particular, had been desperate to see Nala again. It was great to be reunited – even if briefly. Nala, of course, was smothered in kisses by Lydia and slipped straight back into her old routine.

When I bought my ticket for the ferry, I was told that Nala would have to travel in a carrier at all times. Iliana took me to a store where we found a decent-sized one with a large window, so that Nala could always see me. I tried it out at home that night. She didn't like going in it one bit, but there was no other option. Once I got her on the ship, I was sure I'd be able to let her have a sneaky run around the deck.

The ferry was delayed a couple of times because of bad weather, but we finally headed off at the end of March. Iliana, Nick and Lydia were there to wave us away. There weren't so many tears this time; I'd promised to pop back and see them at some point.

We rolled up the ramp and on to the huge ferry late in the evening, ready for a long, overnight trip.

I parked the bike down in the hold, then found a discreet spot on the upper deck where I sneaked Nala out so that she could sit on my shoulder. We watched together as the lights of Piraeus gradually faded into the distance. The port – and all of Athens – was soon a single line of faint, flickering light on the horizon.

It felt like an important moment, as if the first phase of our journey together was drawing to a close and we were ready for a new chapter.

I didn't mind whether my landmarks were measured in miles or kilometres. Either way, each new accomplishment gave me a boost. So, when I worked out that Nala and I had covered more than 1,000 kilometres since we'd found each other back in the mountains of Bosnia, I felt a quiet sense of pride, but also gratitude. We'd been through a lot already, but for all the ups and downs, the surprises and setbacks, I wouldn't have swapped it for the world.

Nala had made my journey so much richer. She was making me a more responsible, more thoughtful person, too. Giving me some purpose. She'd been the best thing to happen to me for a long, long time.

I remembered the words of the Syrian man with whom I'd shared the oranges back at the refugee camp. 'Be a blessing to others and you will be blessed,' he said. If Nala's friendship was my reward for having picked her up on that mountain road, then I was a very blessed man indeed.

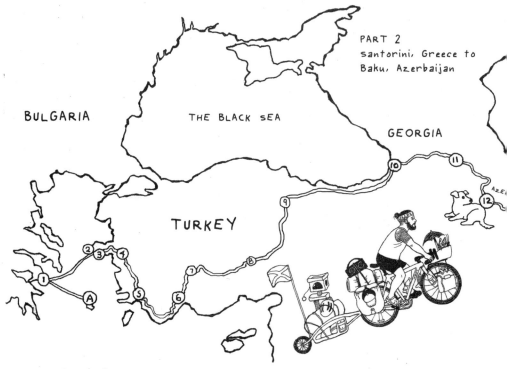

PART 2
santorini, Greece to
Baku, Azerbaijan

BULGARIA

THE BLACK SEA

GEORGIA

TURKEY

MAP 2

A. Santorini, Greece

1. Athens, Greece

2. Chios, Greece

3. Çeşme, Turkey

4. Izmir, Turkey

5. Marmaris, Turkey

6. Antalya, Turkey

7. Köprülü Canyon, Turkey

8. Cappadocia, Turkey

9. Sivas, Turkey

10. Batumi, Georgia

11. Tbilisi, Georgia

12. Ganja, Azerbaijan

B. Baku, Azerbaijan

PART TWO
HIGH ROADS AND LOW ROADS
Greece – Turkey – Georgia – Azerbaijan

10
April Fool

We all get moments when we feel as if our world has been flipped on its head; when something happens that is so totally unexpected, we sense straight away that nothing will be the same again. I guess it was fitting that I had a moment like that on the day Nala and I arrived on Santorini. It was 1st April, after all, and for a while I was convinced it was someone's idea of an April Fool's joke.

Nala and I were recovering from the ferry trip from Piraeus when it happened. The overnight voyage had been uneventful enough. Nala had spent most of the time sleeping, draped as usual across my chest while I lay on the deck in a quiet corner of the ferry. But as the ship began pulling into Santorini, I decided to beat the rush and head down into the hold to reclaim my bike and gear. It was a mistake.

The roar of the ship's massive engines, along with the loud, metallic clanking and grinding of the huge doors as they opened, sent Nala into a wild panic. I'd never seen her behave like this before. The louder the noise, the more she shook and shivered, digging her claws into me.

I felt awful. So as soon as the ship's crew had lowered the safety barrier, I pushed my way through the throng and wheeled the bike down the ramp, clutching Nala close to me. I headed straight over to one of the cafes that stood in the shadow of the island's steep cliffs. She was still shaking like a leaf.

I got her some food and a bowl of water then sat with her, looking out over the famous caldera, the remnants of the huge, sunken volcano that had erupted many million years earlier to form the spectacular island. Nala seemed to be relaxing.

My new boss, Haris, had sent me a note before we left Athens to say his brother was going to pick me up, but there was no sign of anyone in the car park. There had been no phone reception on the ship and my battery was low, so I had switched it off. I pressed it back on again in case Haris was trying to call me.

A waitress arrived with the coffee I'd ordered. My phone had come back to life by now and started making all sorts of pinging noises. I'd received a load of texts, emails and notifications. I panicked a little. I was worried there might be a problem with Haris, but I soon realised it had to be something else. There were too many messages. My notifications from Instagram had gone crazy, too. The little boxes letting me know that someone had liked a post, or started following me, kept coming in waves, pinging each time. My phone resembled a pinball machine.

'What the hell is going on?' I said to Nala.

Most of the emails had the words 'Dodo video' in the subject.

'Hmmm,' I said to myself. The little piece they'd done on us must have gone up on their site. It had obviously attracted a few new followers. I opened up the page and took a look.

I was lucky I didn't spill the boiling hot coffee all over myself.

'What the f★★★?' I said, a little too loudly, drawing a disapproving look from an elderly British tourist.

When I'd left Athens, I'd had around 3,000 followers. I thought it was a very respectable number and I was proud of it. I now had nearly 150,000 people following me. Fifty times as many as the previous day. And that number was rising even as I spoke. People were following me at the rate of a hundred every few seconds. My phone was still pinging like crazy.

I was shellshocked. I couldn't believe it. It must be a mistake, or a joke. Could someone back in Scotland have hacked my account and staged this as an April Fool's stunt? I wouldn't put it past them. It was the sort of prank I'd have loved to pull.

But the more I looked, the more it seemed to be real. Some photos of Nala had been liked tens of thousands of times. A little video I'd made of us back at the beginning of February, as we entered Greece, had more than 150,000 views. It was insane. No one could fake that. Could they?

The explanation came soon enough. One of the emails was from a friend back in Dunbar with a simple message: 'Dean. Have you seen THIS?' She'd included a link to the video on The Dodo's Facebook page. I swore again when it came up. It was titled simply: *Guy Biking Across The World Picks Up A Stray Kitten*. I couldn't watch it. I hated the sound of my own voice. But looking closely at the screen, I could see that it had already been viewed three million times. Three million. That was impossible.

I'd heard about people going viral on the internet and never grasped how it happened. I assumed it was a slow process, or that it involved some big plan that lit the blue touch paper before it exploded. But in this case, there was nothing. No warning. No build up. Just boom!

I looked again at the caldera. It wasn't as big an explosion as that must have been, but it sure felt like one. I already had a suspicion that it was going to reshape the landscape of my life.

I was still processing it all when I heard a car pull up near us, beeping its horn. A young, bearded guy was leaning out of the driver's window, smiling broadly and giving me the thumbs up.

'Dean,' he shouted. I assumed it was Haris's brother.

'Aye, that's me,' I said, gathering my gear.

Nala was a lot calmer now. I placed her on my shoulder and

wheeled my bike and trailer towards the car. The driver had already opened the boot ready to load me up.

'I'm Tony,' he said, extending a hand. 'I work at the kayak school. I'm gonna take you there.'

He looked surprised to see that I'd arrived with a bike and trailer, even more so with a cat in tow as well.

'Nice cat, what's her name?' he asked.

'Nala.'

'Hello, Nala. Welcome to Santorini.'

There wasn't room for the bike in the back of his car, a small VW. So we decided that he'd take my trailer and panniers – to help lighten my load – and I'd cycle behind him.

The climb back up a series of hairpin bends to the top of the cliffs would have been tough going in any weather, but the wind that had whipped up in the past half hour was ridiculous. It felt more like a tornado. At one point, I thought it was going to pick Nala up and send her flying out to sea. The road was really busy, too, which made cycling even more challenging. I was knackered by the time I reached the top.

It gave me a moment to take in the view. Santorini was every bit as magical and beautiful as Nick and Iliana had promised. It was an amazing-looking place, a half-moon shaped chunk of volcanic rock, with pretty, whitewashed villages hanging on the side of its cliffs. The seas that surrounded it were the bluest I've ever seen, even today with white horses breaking all across the bay in the wind. On any other day, I would have happily spent an hour wrapped up in the landscape, but with the wind howling and my head still racing, trying to work out that activity on my phone, I soon kicked on. I wanted to get to our destination.

Tony led me to a large house on the other side of the island, near the village of Akrotiri.

'The kayak staff all stay here,' he said. 'It's just you and me at the moment.'

Tony then took me and Nala to a small beach a ten-minute drive from the house, on the northern coast of the island. It was much more sheltered here, and the winds weren't half as strong as on the southern side. Even now, at the beginning of April, there were already a few holidaymakers with their children, splashing in the surf. I couldn't blame them. The sand may have been volcanic, shades of grey and black, but the sea looked gorgeous. It was a deep, blueish green. I couldn't wait to dive in.

We followed Tony past a few cafes and bars, one or two of which were already getting busy. We then dropped down on to the far end of the beach, a narrow strip of sand hemmed in by powdery, red cliffs about thirty or forty feet high. The beach looked like it could do with a good clean-up. As well as piles of dried-out seaweed, there was all sorts of rubbish washed up on the sand; not only plastic, but driftwood and other odd items swept in by the waves. I'd have my work cut out keeping this place clean.

The kayak base was towards the end of the bay. It was a stone shed, which had been built around a cave that went into the side of the cliff. Inside, Tony opened up its windows and flipped on the strip-lighting. The place was thick with dust and sand that had blown through the gaps in the door and window frames during the winter. It also had a musty smell. It was an Aladdin's cave, filled with kayaks and equipment of all kinds, from helmets and paddles to life vests and ropes.

'Needs a good cleaning,' Tony said. 'And a fresh coat of paint. Rest up today and we can start in the morning.'

We had a couple of drinks together that night, but I turned in early, partly to make sure Nala was okay after her earlier upset, but also to try to get a grip on what was happening online. It had been bugging me all day.

Lying alongside Nala, I scanned my phone again. It was even more insane than earlier. I now had well over 200,000 followers on Instagram. There were hundreds of comments from people all over the world. Another half a million people had watched the

Dodo video. All the other numbers had escalated in the same way. My email was filled with dozens and dozens of messages from all sorts of people, but also companies. I saw the name Netflix on one of them. Someone there was suggesting they come and film me. I deleted the message. I couldn't get my head around that right now.

There was also a slew of enquiries from news agencies and newspapers. I replied to a couple, one from the *Daily Mail,* which was the paper my parents read back home, and another from the *Washington Post* in America. Even I'd heard of that one, although why such a serious newspaper would be contacting me was beyond me. I agreed to have a chat with them both in the coming days.

At times like this, when I feel overwhelmed by things, my natural reaction is to switch off and ignore whatever's causing it. Some people might call this tactic burying my head in the sand, but I prefer to think of it as taking a breather. Regrouping. I knew that if I started going through all this in detail, I'd be opening a massive can of worms. It would send me into a spin; it wouldn't be good for me.

So once I'd updated my Instagram – being sure to thank and welcome my new followers, of course – I put the phone to one side and tried to get some sleep. I knew I had a busy day ahead of me.

Tony and I headed down to the kayak base bright and early the next morning. He'd picked up some paint and brushes, but before we could start decorating, we needed to clear the place up. Every corner seemed to be stacked with equipment or boxes. We put on some face-masks to protect us from the dust and got stuck in.

Tony seemed to be a decent guy; he was easy-going and liked a laugh. We'd not spoken much the previous night, as there had been too many other people around. But the moment he switched on some music on his sound system inside the cave, I knew we were going to get on like a house on fire. He had exactly the same taste as me. In particular, he loved a house DJ called Solomun, whom

I played a lot. We made light work of the clearing up, singing and dancing as we disposed of rubbish and brushed the inside of the cave clean. It was so much fun that it didn't seem like work at all.

Nala was in her element, too, skipping around in the small cluster of rocks a few yards in front of the cave, chasing and dodging the waves as they rushed in and out. I'd stuck a few empty boxes outside, knowing she'd want to play with them. With the doors of the cave wide open, I kept an eye on her at all times.

'So what's the story with the cat?' Tony asked me, as we painted away.

'It's a long one,' I said. 'Short version is that I rescued her back in Bosnia. Four months ago. Been with me ever since.'

'Sure she didn't rescue you?' he said, with a wink. 'I've seen the way you are with her. Looks like you found your soul mate.'

I smiled. He was the first person to point out what must have been blindingly obvious to anyone. Nala and I now came as a pair. At the end of the afternoon, Tony and I wrapped up and admired our handiwork. The place looked one thousand per cent better.

'Shall we go somewhere for a beer?' Tony said.

'As long as they take cats,' I replied.

He just laughed.

We walked a few hundred yards along the beach to a bar, then sat in a corner, overlooking the sea. The sun had set on the other side of the island and the sky was glowing red. It was shaping up to be a perfect evening.

Tony and I had taken a sip from our beers when a group of local girls walked past. Nala was standing on a wall next to me, looking out to sea, and made a little chirping sound at them. One of the girls saw us and smiled. She then stopped in her tracks, her face locked in an expression that I could describe only as shock. She was soon pointing and muttering to her mates.

One of them could speak English and came over.

'Are you the guy from Instagram? The one who rescued the cat?' she said.

'Aye,' I said after a moment or two.

I was too gobsmacked to say anything else.

'My friend follows you. Can we take a photo?' the girls said.

'Sure.'

Nala and I were soon posing for a series of snaps. It seemed to make the girls' night; they ran off, giggling away as they looked at the photos on their phones.

For a moment Tony fixed me with a look.

'Okay. What the hell was that about?' he asked.

I hadn't planned on mentioning it, but I didn't have any option now. I showed him the Dodo video and my Instagram page.

He whistled loudly.

'So we've got celebrities working for us this summer,' he laughed.

I'd never thought of it in those terms. And I didn't really want to now.

'Nah, just a fluke that she'd seen it. Won't happen again,' I said.

'Of course not,' he replied, with a knowing smile.

He was right. And we both knew it. This was only the beginning.

We spent the next day or so finishing off preparations for the reopening of the kayak school, a couple of days later.

The best bit came when Tony suggested we take the kayaks out for a test.

'We need to know they still float,' he joked. 'But I also need to show you where you will be taking the tourists.'

I wanted to take Nala out at some point, but wasn't going to do that right away. First, I needed to familiarise myself with the waters and the coastline. I wasn't going to do anything daft.

So I left her inside the base with plenty of food and water while Tony and I paddled our way off around the coast. We'd strapped a boombox on the back of one of the kayaks so that we could play music. For a couple of hours, we worked our way along the coastline. I'd been kayaking since I was a teenager and found it straightforward.

'Looks like you really have been in a boat before,' Tony shouted, as we negotiated one tricky little section where the winds had come up.

Tony knew the waters like the back of his hand and guided me through the route we'd be taking the kayak trips, pointing out the places where I'd need to watch out for changing currents, or sudden, strong winds. He then led me into a little cove, where we'd pull in for a packed lunch each day. He warned me that landing there could be difficult in the strong winds that regularly blew around the Aegean.

For a while we glided along, listening to music and chatting. *If this is the way I'm going to spend the rest of my summer*, I thought, *I may never leave this place.* It was perfect. I'd found my dream job.

The following day a new member of the team turned up, a nice Slovenian guy called David. He was also going to be living at the house. Haris popped in briefly to say hello as well. He was a little older than Tony and spoke less English, but he shared his brother's easy sense of humour. It was clear that he'd be only an occasional presence. He seemed to have other businesses on another island and left Tony to manage the day-to-day running of the kayak base here on Santorini.

With the team assembling, Tony walked us through what lay ahead for us in the coming season, scribbling on a whiteboard while David and I made notes. We'd be running one or two expeditions each day, mainly for people who had some experience of kayaking. Some would start from the base and others from the other side of the island, inside the caldera.

As far as the timings were concerned, the first visitors would arrive around nine in the morning. We'd take them through an introduction, get them kitted out and into their kayaks and set off on the first of our two- to three-hour trips. We'd pull into the cove halfway around, where we'd give them a packed lunch. The day's second trip would usually be in the late afternoon, to watch the famous sunsets that drew people to Santorini. Apparently, they were spectacular. That too would take a few hours and we'd arrive back at base early in the evening.

My heart sank as he spoke. This was going to be a big problem for me. I wasn't going to leave Nala back in the house from early in the morning until the evening. I'd be too far away from her. I didn't say anything to Tony: he'd hired me to work on all his expeditions so I couldn't make myself unavailable for some of them. Luckily, I soon came up with a solution.

The following day, when we went down to the cave, I quietly slipped my rucksack and some camping gear into the back of the car. Inside the cave, I pulled out my hammock and gave it a good dusting down. I hung up my clothes in a corner of the shed and started arranging Nala's stuff nearby.

'What are you doing?' Tony said, when he saw me. 'Hiding from the paparazzi?'

'Thought this place could do with a guard cat,' I smiled back.

When I explained my concerns about leaving Nala up at the house, he was absolutely fine. He even offered to make sure that whichever members of the team were left at base looked after Nala while I was out at sea.

'We can make it work, I'm sure,' he said.

It made me feel a million times better.

Our makeshift home didn't need much setting up. I rigged up the hammock, then popped down to one of the shops on the road above the beach and stocked up on some coffee, pasta and food for Nala. We were sorted. That first night in the cave felt like we'd moved into our first apartment together. I rustled up some food for us both, sat watching the stars for a while, then turned in, ready for an early start the next day. After a couple of days of windy weather, tomorrow was due to be clear. The first tourist party of the season were booked in and my job would start for real.

As Nala snoozed on my chest, I caught up with my emails. We

now had more than 300,000 followers, while the Dodo video had been watched five million times. I'd posted a photo of Nala playing on the beach and it had been liked 100,000 times in less than twenty-four hours.

It was ironic. Everything now seemed to have loads of zeroes attached, which would have been a dream come true for many people. And yet I had zero idea how I was going to handle what was happening to me. This had not been part of my plan.

The one thing I managed to do was speak to a couple of the journalists who'd contacted me. The first was a nice woman back in the UK, who was going to write a piece for the *Daily Mail*. I then spoke to a very serious-sounding guy from the *Washington Post*. I couldn't believe he was talking to me. Didn't he have wars or big political stories to write about?

It still felt weird to be talking about myself, but I retold the same tale, polishing it up each time. How I'd found Nala on the mountain in Bosnia. How I'd sneaked her through the borders. How we'd bonded into a team. Scotty and Kirk. Neither of them could tell me when their articles were going to be published, but I was fairly sure they wouldn't be; that this was all a flash in the pan.

I couldn't deny the interest on Instagram, though. As the numbers kept growing, so did the level of praise being heaped on me in the comments and messages sections of the page. People were making me out to be some kind of modern-day saint. I would read their contributions and shake my head. I thought it was daft. I'd done what any decent person would have done. There was nothing special about me, far from it. At times it made my head spin: it began to feel too much, that I'd set something in motion there was no way I could live up to.

Thankfully, I had Nala there to ease my worries.

She was fast asleep, splayed out across my chest, her legs turned up into the air like a dead fly. It always amazed me how she could sleep anywhere – draped across the handlebars of my bike, up a tree; she could adapt to anything. She has the right idea, I told myself, as I put my phone to one side. I have to get used to this new situation. Work out what to do. Find a way – like Nala always does – to be comfortable with whatever lies ahead. We'll be fine. I just have to give it time.

The sound of the waves gently lapping outside soon sent me to sleep.

11
Nurse Nala

Nala's habit of waking me at the crack of dawn wasn't all bad. Next day, as I stumbled bleary-eyed to the doors of the cave and let her out to do her toilet, I looked out on to the most glorious sunrise. I also saw that we had the beach completely to ourselves. I thought, *I might get used to this.*

While Nala darted after the waves and sniffed at the fresh seaweed that had washed up overnight, I hopped down onto the beach and walked a little way, breathing in the sea air and enjoying the almost eerie calm. It was a little after six-thirty. The only sound was the gentle breaking of the waves and a dog barking somewhere in the far distance. It reminded me of walking the deserted sands back home in Dunbar in wintertime, something I'd always loved. It was already twenty degrees warmer here, of course.

The winds of the previous day had eased off, but tell-tale flecks of white were still forming a short distance off the shore. There might be some rough waters to negotiate when I went out with my first tour party in a few hours' time, but I wasn't too worried. I was looking forward to getting stuck into my new job.

By eight o'clock or so, the beach had got a little busier with joggers and early-morning swimmers, so I began getting ready for the day ahead. I was sorting out my wetsuit when my phone made a pinging sound. My heart sank the instant I read the text message.

I looked out through the open doors at Nala and sighed. She was still enjoying herself, jumping around the rocks that had become her favourite playground. Poor thing. It seemed we'd both have some choppy waters to face.

The text was from a vet on the island, whom I'd contacted soon after arriving. He was offering me an appointment for Nala to be spayed the next day.

I found it hard to believe she'd turned six months the day after we landed in Santorini. She'd grown physically, of course. But she was still a lithe, athletic creature. Her waist was so thin I could wrap my hand around it. She was a baby in temperament, too – well, as far as I was concerned anyway. Given the chance, she'd spend all day chasing a toy on the end of a string or a beam of light on a wall. All in all, she seemed too young, too innocent to need such a drastic procedure.

I knew it was due, but I'd tried my best to postpone it. As well as contacting the vet, I'd called a woman called Lucia, who ran Sterila, a charity on Santorini that worked to help the island's large population of stray cats. Apparently stray cats were a big problem and I was interested in helping while I was staying in Akrotiri. Before asking how I could contribute, I asked her advice about Nala. I also had an email exchange with Sheme, the vet who had helped me back in Albania. He'd been keeping me up to date on Balou, who had now left Tirana and was living happily back in the UK.

My question to all three of them was the same. Is it really necessary to spay her? Won't she be okay travelling with me? They were all fairly blunt about it. Without the operation, she'd be vulnerable to cancers, tumours and infections. By not doing it, I risked shortening her life. On top of that, the procedure would also reduce the chances of her producing some little Nalas, which was a pretty persuasive argument. Carrying a litter of kittens around on my bike was not on the cards.

I'd learned my lessons by now about ducking big decisions when it came to Nala's health. So, with a slightly heavy heart, I'd bitten the bullet and asked the vet to contact me if there were any free appointments. I'd been secretly hoping his diary was full, but his text this morning put an end to that idea. He asked me to drop her off at his surgery first thing the next day.

By the time Tony and David arrived at the base a little before nine, I was glad to have some distractions. I knew it was a routine procedure, but it didn't stop me worrying. I was her dad and she was my little girl, after all.

It was the first full day of the season and we were welcoming the first paying customers. Tony reckoned that the weather forecast was reasonably good.

As this was my start day as an official member of his team, he wanted me to do the trip as if I was a tourist, travelling in one of the double kayaks that the customers used. He'd be alongside us in a single seater. This would give me an understanding of their experience. Tony was also interested in my feedback on the trip. Was it value for money? What were the best and worst bits? What could he do to give the customer the best possible day out on the water?

The customers arrived a little after nine. The party was small, only three American guys who were staying over in the upmarket village of Oia, on the west of the island. One of the trio was absolutely enormous. He was way over six feet tall and must have weighed at least twenty stone.

Tony and I exchanged looks when he was walking down the beach.

'I think we're gonna need a bigger boat,' I half-joked.

The three of them were fairly experienced kayakers, as it turned out. The only problem was the big guy's size. He was barely able to fit our biggest life vest around himself. I got in the back of his double kayak so that I could keep an eye on him, but because of his weight, the boat was unstable from the moment we hit the water. I considered suggesting we switch to single kayaks, which customers weren't normally allowed to use, but I thought better of it. I was the new boy, so I couldn't cause a scene on my first day at work.

As we headed out, staying close to the coast, I did my best to stabilise the boat. But the kayak was regularly too low in the water. I guessed this was going to be hard going. I wasn't wrong.

Tony had told me that the waters around Santorini were unpredictable.

'Always expect the unexpected,' he said.

An hour or so into our trip, I found out exactly what he meant.

What had been a gentle breeze suddenly became a wind that was threatening to blow off the baseball hat I'd worn to fend off the sun. Before we knew it, the sea was producing waves two to three feet in height.

Tony had been concentrating on the two guys in the other kayak. With the wind stiffening, he was trying to lead them nearer to shore. But then, out of nowhere, a series of waves hit them, sending the two guys into the water. We knew they were safe – they were both wearing life vests – but Tony was forced to go to their aid, holding their boat while they clung on to the side.

As all this was happening, my boat was slowly sinking. The waves were coming straight over the top of us and filling up the kayak. I'm not a panicker. I grew up in water and am a strong swimmer; I've done some lifesaving training, too. But I could tell the guy was getting anxious. He didn't like the worsening conditions and was asking me to take him in to shore.

We were closing in on a small cove when a wave sent us over. I lost my grip on the kayak and was soon watching it slide away from me, further out to sea.

I knew Tony was a skilful seaman, but his reaction was amazing.

While still holding on to the other kayak with one hand, he got hold of ours with the other. Somehow, he then pushed his little convoy back towards me. I grabbed the kayak and swam on to the beach, making sure the big guy was alongside me. He was.

That wasn't the end of our problems. I checked out the cove and saw that we were hemmed in by sheer cliffs on all sides, so there wasn't a way out on foot. There was no option but for the two of us to get ourselves back into the kayak. The guy was a little calmer, now that we were closer to shore.

We soon found a bigger bay with a pathway up onto the top of the cliffs. Tony landed a few minutes later with the others in tow and everyone seemed fine, though relieved to be back on dry land.

The wind was even stiffer by now, so we all agreed to call it a day and carry the kayaks back on foot. By the time we reached base, we were all shattered.

The customers were happy enough. They seemed to have regarded it as a bit of an adventure.

'Reminded me of white water rafting back in Colorado,' the big guy joked.

For my part, I certainly marked it down as useful experience. It was part and parcel of kayaking. I'd face these conditions again this summer, I was sure.

Back inside the cave, I found Nala fast asleep on my jumper, oblivious to the dramas of my day. David told me that she'd been as good as gold and hadn't even tried to leave the cave while I was away, which gave me some comfort. It was good to know she could be left on her own for a few hours with the team back at base.

With Tony and David, I hosed down the kayaks and the equipment, so they were clean for the following morning. It took an age to clear out every bit of sand, which seemed to gather in every crevice. I then hauled the kayaks up into the racks, where they were stacked methodically. It took longer than I imagined, but I'd have to get used to it. This was what every day for the next few months was going to look like.

With no other trips booked in for the day, I had most of the afternoon and evening to myself. I spent it walking the beach with Nala and generally chilling out. Nala, of course, was completely unaware of what lay ahead for her the next day, but I couldn't get it out of my mind. I wasn't able to stop myself looking up the spaying operation on my phone. Too much information isn't good for you, so as I caught myself starting to view the graphic details, I quickly gave myself a ticking off. *Stop it.*

The following morning, I let Nala have an extra ten minutes running around on the beach before slipping her into the cat carrier and driving about three miles with Tony up to the surgery in Fira.

The staff at the surgery were very professional and talked me through the whole process, putting me at ease as much as possible. She'd have the procedure later that day, then need time to come round from the anaesthetic and be checked over. So I was told to expect a phone call twelve hours later, around eight o'clock that night. I was also warned that I might have to wait even later if she was still drowsy.

I gave Nala a nuzzle and left her with a nurse. I knew she was in safe hands, but walking out of the door, I still felt like a total traitor. I couldn't look her in the eye. I set off back to the kayak base with a heavy heart. I knew this kind of operation happened every day without problems and that I shouldn't be worrying. But the bond between Nala and me had become so strong, I couldn't help myself. At least I had a day of kayaking to distract me.

The second day turned out to be a lot less dramatic than the first. Tony let me go out in a single kayak as a supervisor, which made life much easier for me. The party we had was a mix of nationalities, some Brits and Americans, but a couple of Germans too. There were eight of us in all. The weather had eased off a little, but we were extra careful, especially around the exposed sections. I was pleased when we slid the boats back onto the beach with a happy bunch of customers. It felt as if the season was up and running after the false start of the previous day.

I spent the rest of the day trying to keep myself busy. As I often did now, I spent half an hour doing a beach clean. I never knew what I was going to find washed up or discarded on our little stretch of coast. One day I found a pair of mismatched, black and white 'Crocs' – moulded plastic clogs. I hung on to them, since I found them really comfortable and figured they'd be handy when I was back on the road on the bike.

But as I loaded up a black bin liner, I couldn't stop looking at my phone. The clock seemed to be moving in slow motion. It felt as if it took an age to get to six, then seven, then eight o'clock, but there was still no call. It went to quarter past, then half past. My mind was going at one million miles an hour.

Had there been complications?

The phone eventually rang at 8.45 p.m.

'You can come and collect your cat now,' the voice said matter-of-factly. Tony wasn't around to give me a lift, so I caught a taxi up to the clinic and asked him to wait.

When the buzzer let me into the clinic, I sprinted into the recovery room. Nala was still really groggy. She barely recognised me. I thanked the vet, then wrapped her in a little blanket and drove back with her sitting on my lap. The sense of relief I felt to have her back safe and well was ridiculous. *How can you get so worked up about a wee cat?*

The vet had warned me that she might be a little nauseous for a while. So back at the cave I decided that I wouldn't sleep in the hammock. I'd lie on the floor with her, in case she was disoriented or sick when she came round. That way, I'd be the first thing she saw when she opened her eyes. I didn't want her to worry that she was on her own.

It proved the right move. When I lay next to her, she instinctively cuddled up. It seemed to calm her and she soon settled back into a deep sleep. I couldn't sleep a wink, of course, which was just as well. She was sick twice, vomiting on the edge of the carpet where we were sleeping. I cleared up both times, then sat watching her as she settled again. I knew it was doing her good; she was getting the anaesthetic out of her system. But it didn't help me relax. I was so on edge I reacted to every twitch and sound she made.

I finally fell asleep in the small hours, but wasn't allowed to doze for long. When I woke up, I found Nala standing next to me. She'd clearly shaken off the anaesthetic, because she was now trying to pick out the stitches on her lower abdomen with her teeth.

'Nala, no!' I said, jumping up.

The little hiss she gave me spoke volumes. *They're my stitches and I'll pick them if I want.*

I knew I had to stop her, so I improvised a makeshift cone from one of the little nylon bucket hats that we had in the kayak shed. I cut the top off the hat, then threaded it on to Nala's neck, upside down, so that the wide rim stopped her biting into anything. It did the job. Nala grizzled and griped for half an hour or so, but she soon gave up and went back to sleep. We were woken by Tony, who'd arrived to start preparing for that day's arrivals. His eyes nearly jumped out of his head when he saw Nala.

'What the hell have you done with our hat?' he said.

'Take it out of my wages if you like,' I said.

He laughed.

It took Nala a few days to recover. She lay around the cave, finding spots that were warm or cool enough for her at different times of the day. I kept her hydrated and well fed, but as I watched her, I remembered something I'd read once about how cats are great at self-medicating, at healing themselves when they aren't well. Normally Nala could have spent ten hours a day running around the beach, chasing waves and playing in the rocks. But she seemed to know intuitively that she needed to rest in order to heal. She knuckled down and carried on with the job of getting better.

As luck would have it, the weather turned too windy to go out in the kayaks, so I was around a little more to keep an eye on her. I also caught up on my Instagram; our ups and downs provided me with plenty to write and post about. Our followers seemed genuinely interested in what we were doing, but especially in how Nala was coping. So I posted plenty of photos of her looking happy and healthy.

People's interest in us seemed to be growing. By now, both the *Daily Mail* and the *Washington Post* had published their pieces, which had led to other publications around the world picking up on the story. As a result, we started to get messages from tourists who were in the Greek islands, asking whether they could come and meet us.

A few days after Nala's operation, we were visited by a young Swedish girl and her parents. They had come down to Akrotiri and found us at the kayak base. It caught me by surprise, but they were lovely people and we spent ten minutes chatting and taking photos. Nala was almost recovered from her op at that point and was in good form. The little girl was beside herself with excitement to meet her. The dad even offered to buy me a beer at the bar along the beach.

'Don't tell me, just another fluke,' Tony joked, when he walked past and saw the family taking photos with Nala and me.

It wasn't long before we were heading back up to the vets in Fira.

During the spaying process, I'd also asked about the Titer test for rabies, which I knew was required for Nala's passport. We'd not get very far on our travels without it. The vet recommended doing it as soon as possible, since it took a while for the red tape to be sorted. The test was essentially a blood sample, but the procedure could take a lot out of her physically, so he suggested waiting a couple of weeks for her to recover from the spaying. Especially if she needed sedating again, which was a possibility.

A little more than a fortnight after the last trip, I packed Nala into the carrier and drove her back up to the clinic.

I sensed she knew where she was headed and I felt bad for putting her through so much in a short space of time. I kept telling myself that I was doing the right thing. I needed to get all these procedures out of the way now. Then she'd be free to enjoy the next year or more without anything major. It didn't make me feel much better.

We arrived at the clinic early. This time I'd decided to stay with her.

I'd been hoping that she wouldn't need to be sedated again. No chance. The minute the vet produced his syringe, Nala went into full lioness mode. She snarled and slashed her paws at him, almost drawing blood. I helped him out and picked her up, holding her close to me and calming her down as he sedated her. She was soon asleep, but I felt sorry for the poor thing.

The dose of sedative wasn't as strong as her previous one and, although she was still groggy and sleepy, Nala was ready to leave about an hour later. By then the vet had managed to freak me out not once but twice.

First of all, he found something that he didn't like in Nala's passport. He started showing it to the nurse who was assisting him, shaking his head and looking grim-faced.

'The vet back in Albania didn't stamp the passport properly,' he said.

'What does that mean?'

'Well, it might mean all her inoculations are invalid. She'll have to have them again.'

My face must have been a picture.

'No way,' I said.

But before I could get too upset, the nurse and the vet were deep in conversation. Things soon calmed down.

'Don't worry,' the nurse said, smiling. 'We have the name of the vet. We can contact him. He can confirm it was all done correctly and send new paperwork perhaps.'

'He'd better,' I said, my nerves settling down a little.

I'd barely recovered from that shock, when another member of staff appeared holding one of Nala's vials of blood up to the light, looking at it strangely with the vet.

'Something wrong with it?' I asked.

'It's very cloudy. Quite unusual. Could be because she was agitated,' the vet answered. 'The test will pick up anything they need to worry about.'

'And how long will it take for the results?'

'About a month. It needs to go to Athens.'

Perfect. As if I didn't have enough to worry about.

The staff at the surgery were all brilliant, but I was glad to get out of the place and head back down to the kayak base, where I put Nala in a little cardboard box with a blanket and let her sleep for the rest of the day.

It knocked the stuffing out of her for the next couple of days, but she always managed to find the right corner of the cave to get on with her recuperation.

Seeing anyone in a vulnerable state in a hospital or surgery makes you protective of them at the best of times. It's only natural and, maybe because I felt so guilty at having put Nala through so much, I found myself being very paternal towards her. I bought her the best food I could find and spent more time than usual lying with her, stroking her head and neck while she lay there purring.

The bad weather passed quickly and I began going out on the kayaks again. However, I wasn't keen on the idea of leaving Nala. It wasn't that I didn't trust Tony and the guys at the base. I did, but it was more the idea of her missing me. I didn't want to be stuck out at sea if she had a problem.

So, in the wake of her ops, I made a decision and bought Nala a lifejacket. She was my co-pilot on the road; she could be my co-pilot at sea as well. Once she'd recovered, I tried it on her and she looked really cool in her bright yellow jacket. The photograph of her wearing it was an instant hit with her admirers on Instagram.

I waited for a really calm day before taking her out for her first trip. She sat in the cockpit of the kayak, like she did in the pouch of the bike, with her body safe and warm inside and her head poking out, taking in everything. She was fascinated by the views, drawn by the music pumping out from the tavernas along the beach.

Cats aren't naturally fond of water, so to ensure she was at ease, I also took her out on a paddle board. That really threw her. It was flat, so there was nowhere for her to hide away, and at first she did look unsettled. She turned to me once as if to say: *What the heck is going on here?* But once we'd been floating and stable for a wee while, she started walking around the board. She took to it like, well, a cat to water. It made me feel better. I didn't plan on taking her out every day; that wasn't practical. But I had a Plan B up my sleeve.

Unfortunately, in the time I'd been wrapping myself up in Nala's welfare, I'd been forgetting to look after myself so well.

Soon after Nala's injections, I had a bit of a disaster and lost my GoPro, a powerful little, cube-shaped video camera that was especially good for filming while I was cycling. I'd bought it before we left Dunbar and had been using it ever since, often wearing it attached to a strap on my forehead, and I took it on an expedition with Tony. The weather wasn't great when we set off and it deteriorated even more when we were out at sea. We hit some heavy waves to the east of our base. I thought I was doing fine, but a swell took me up and flipped me over.

I got myself back into the kayak when, completely out of the blue, a six-foot wave hammered into me from behind, sucking me out of the kayak again. I didn't realise it at first, but when I checked my forehead, I found the strap holding the GoPro had been swept clean off.

I was annoyed, but not devastated. The camera was very sturdy and came in a sealed, waterproof casing. I expected I'd be able to find it on the seabed. At the point where I'd lost it, the waters weren't that deep; if the sea was calm, I should be able to recover it. But I was kidding myself. I spent about two days scouring the seabed, but to no avail.

As if that wasn't bad enough, I took Nala out on one final search expedition. I was so concerned with making sure she was all right that I forgot to go through my usual routine. My Scottish complexion doesn't agree with the fierce Greek heat, so I have to be careful to smear myself with factor five thousand sun cream on my legs and arms. But for once, I was not careful enough to cover my back and neck before going snorkelling. I spent about an hour face down in the water with both of them exposed.

It hit me that evening. It was as if I'd been run over by a bus. The nausea and dizziness were ridiculous, and I was even sick a couple of times. My body felt drained of all energy as I lay there in my hammock. I recognised what it was: I had it before when I'd been on holiday in Thailand. I'd got sunstroke.

Tony spotted how sick I was and loaded me up with water and some medication.

'Take a couple of days off,' he said, drawing grunts of protest from me.

When I got up the next morning and started sorting some of the gear, he grabbed me by the arm and physically led me back to my hammock.

'Don't even think about doing that again until you're recovered.'

Reluctantly, I went back to my bed. Our little home was turning into a sick bay, only this time Nala was the nurse and I was the patient, or that's how it seemed. Nala was a star. It was as if the roles had been reversed. Rather than sleeping around the cave, as she often did, she remained glued to my side, curled up close to me, purring and licking my face every now and again. It was as if she knew that it was me and not her who was now not one hundred per cent. It was her turn to keep a watchful eye on me. I appreciated her company, especially as I wasn't out of the woods yet.

Soon after I'd shaken off the worst of the fever, my left leg grew inflamed and began itching like crazy. At first, I put it down to the sunstroke or an insect bite that had turned septic. I took some painkillers and antihistamine tablets, but they did little good. Then my skin began to turn a grim reddish colour. After a day or so, it got to the point where it had hardened up and I couldn't straighten my leg. Putting weight on it was also out of the question, and I started to feel like Long John Silver again – this time when I started hopping around like him, minus his crutch. I raided the first aid kit at the base, but nothing seemed to help.

Tony came in the next day, took one look at my leg and read me the riot act.

'You can't kayak like that,' he said. 'You need to go to hospital.'

As soon as he'd finished closing up the cave that evening, he drove me there.

I didn't think it was worth the bother, but the nurses seemed to disagree. Before I knew it, I'd been seen by a doctor, stuck in a little cubicle and attached to an antibiotic drip.

I was told that the leg wasn't infected, but I'd probably had an allergic reaction to something. I'd also been seriously dehydrated. After a few hours lying on my back, bored, they discharged me and sent me home with some more antibiotics and instructions to take it easy the next few days.

I had little intention of listening to them. I'd lost too much time off work already. I was already telling myself that I'd be back in the kayak the next morning.

Nurse Nala had other ideas.

They say that cats can sniff illnesses, sometimes more accurately than machines. I've read stories of cats spotting when their owners are about to have epileptic fits, for instance. As soon as I got back to the cave and lay in my hammock, Nala came bounding over and sprang up to lie with me. She leaned in close to me and purred quietly away. It was as if she knew I wasn't well and needed some TLC.

I've always been a terrible patient. I was the same as a kid back in Scotland.

I'd ignore cuts and bruises. Think nothing of a cold or a dose of the flu. When I played rugby, it would have taken a broken leg to get me off the pitch. It was mostly macho nonsense, I'll admit. A facade, trying to look like the tough guy, as if I was indestructible. It was the same back in Bosnia, when I busted my leg jumping off the bridge in Mostar. I should have waited for it to heal properly, but I didn't. I was lucky to get away with it. I could have been walking with a limp for the rest of my life.

As I lay there in my hammock, turning things over in my mind with Nala beside me, a penny finally dropped. Every instinct within me wanted to jump out of bed and get out there. I didn't like shirking work and I was terrible at doing nothing. But I wasn't going to be any use to Nala if I was hobbling around on crutches. Tony certainly wouldn't let me anywhere near a kayak if I was in that state. So what was ignoring the doctor's orders going to achieve? I might only make things worse.

I remembered how Nala had handled her recoveries. How she'd forgotten everything else and focused totally on getting well again. It felt silly, but I found myself thinking that perhaps I should take a leaf out of her book and give myself time to heal properly. I should, for once, listen to my body.

Almost at that precise moment, Tony popped his head round the corner.

'How's the patient?' he asked.

'Think I'll need a couple more days to get straight,' I said.

He looked surprised.

'You sure? Not like you.'

'Aye, a one-legged kayaker is no good to you.'

'Good,' he said, still slightly disbelieving. 'So, who persuaded you. The doctor?'

I shrugged.

'No,' I said, avoiding eye contact. 'Just seems like the right thing to do.'

I still had some pride left. I wasn't going to admit that it had taken a six-month-old kitten to knock some common sense into me.

12
The Spiderman of Santorini

By the time the summer season had got into its stride, Nala and I felt fully settled on Santorini. It was a simple life on the beach in Akrotiri, but it suited us down to the ground. My routine was a million miles from the grim, six-till-four grind of Scotland. I worked hard, but played hard too. It was a party island and I burned the candle at both ends every now and again. Nala also seemed entirely content living on the edge of the sea. Like me, her only real problem was the heat. By now the temperatures were climbing into the eighties and even the nineties during the afternoons.

Too much exposure was as bad for cats as it was for humans. After my experience with sunstroke, I wasn't going to risk Nala getting overheated. So, following the advice I'd read online, I started smearing her ears and nose with special cat-friendly sun cream. Apparently the bare skin there was the most easily and badly burned. I also decided against taking her out on the water with me too often. It was fine on short trips, but she couldn't join me for three-hour expeditions, much as I would love to have her alongside me.

It wasn't ideal. Taking her with me put my mind at rest. Without her at my side, I spent my time out at sea fretting constantly about what was going on back at home. How busy was the beach today? Who was hanging around the base? Was Nala safe inside? It wasn't helped by what had been happening on Instagram.

We now had close to half a million followers. People were following us all over the world, in countries from Canada and the USA to Poland and Brazil. A lot more people were also following us here in the Greek islands, where they were on holiday. As a result, the trickle of messages asking to come and meet us had turned into a torrent. There were one or two every day. I tried to accommodate everyone, but I couldn't always make people happy. A couple of times people turned up unannounced as I was about to head out with the kayaks and I had to disappoint them. I also arrived back at the base one day after taking out a tour party, to find we'd had visitors asking to see us. They had also left crestfallen, without seeing Nala.

It made me feel bad. People had gone to the trouble of looking us up and travelling over to Akrotiri, often from the other side of the island. But I had to be fair to Tony. He and Haris were employing me to run kayak trips, and I'd already had to take time off because of my heatstroke and leg. It wasn't fair on Nala, either.

Those that did get to meet Nala were lovely. They'd spend some time talking and taking photos, then leave with big grins on their faces. A few invited me for a drink afterwards. A group of girls from the UK and Australia even dragged me out on the town. We ended up in a tattoo parlour getting matching pineapples inked on to our ankles. It was quite a night.

It beggared belief how many people had been touched by our little story. In addition to emails and private messages from more publishers, agents and journalists, parcels were arriving at the local post office as well. They were coming from around the world – but mostly the US – addressed to 'Dean and Nala, Santorini'. They contained all sorts of gifts, from posh food and toy mice to bells, harnesses and catnip – every cat treat known to humanity.

The problem was that when they arrived in Greece, they did so with tax still to pay. So I had to fork out each time for them to release the packages. It was a game of chance; I'd never know whether it was worth it or not. I paid fifty euros once for a small box that contained a stick with a feather on the end of it.

I couldn't complain. People were being incredibly generous and their hearts were in the right place. But I knew I had to put a stop to it, so I posted on Instagram, telling people to donate their gifts to their local animal shelters instead. I couldn't keep everything, especially when I went back on the road. If I did, I'd end up travelling around the world in a mobile pet shop.

People's reaction to us was genuinely overwhelming, but I didn't want to lose sight of my main purpose. I wanted to use our new profile – our 'influence' – to do something useful. To do some good. Thankfully, some opportunities began to appear that May.

Beneath the picture-postcard, perfect facade, Santorini had a dark side. It was during a visit to Fira on a day off that I first noticed the island's population of donkeys and mules. I was taking in the view of the caldera, and the tourist boats coming in and out at the foot of Fira's huge cliffs, when I saw a number of people arriving up in the village on the back of donkeys. I could tell the poor animals were being driven hard. One arrived with a really big guy sitting on its back; it was bathed in sweat and panting hard. I thought it would buckle under the weight at one point.

I didn't understand why the guy had put the poor creature through it. There was a cable car that could have taken him up. He also had two working legs, as far as I could see.

After that, I noticed other mules around the island, some pulling carts for farmers, but one or two older ones standing alone in fields or wandering the roads. I read about this online and found out that the 'taxi' donkeys worked from dawn 'til dusk, seven days a week, during the holiday season, so they developed problems with their joints, legs and backs. By the time their working days were over, they were in a terrible physical state. Unfortunately, their owners had little interest in their welfare. Many simply abandoned them.

So when Lucia from Sterila offered to take me on a visit to another charity on the island, devoted to looking after ageing donkeys and mules, I jumped at the chance.

SAWA, or the Santorini Animal Welfare Association, was based near Fira. As well as looking after ageing donkeys, they also had cages full of stray and abandoned dogs. They had a whole pen full of pointers, a breed I've loved since my family had had one, named Teal, when I was a kid back in Scotland. I wasn't able to resist and spent a few hours playing with half a dozen of the pointers, while chatting with Lucia and some of the volunteers.

One of them was a young Greek woman from Athens. She told me the charity was established at the start of the 1990s by her boss, a lady called Christina. Cats, dogs and donkeys were their main concern, but they'd recently had a couple of pigs and even taken in abandoned farm animals. Their aim was to neuter, spay and vaccinate all strays and treat any injuries or diseases, before rehoming the animals elsewhere in Greece or overseas. The charity's biggest challenge was the Greek government, who didn't see animal welfare as a huge priority. They wouldn't give SAWA a penny in aid. Their main shelter had taken a real pounding during the worst of the winter and was in dire need of funds.

I was struck by the work they were doing and admired their determination. So when I got back to the base that evening, I put up a post on Instagram explaining the plight of the donkeys and asking people to help patch up the shelter. I didn't expect too much and, after I posted, I spent time with Nala in the evening, walking and clearing the beach. It was only when I got a message from Christina the following morning that I saw the reaction online. It had been instant – and amazing. The appeal had raised several thousand euros, easily enough to start work on fixing her shelter.

Christina didn't know what to say at first.

The response knocked me for six as well. I'd seen this on a

smaller scale when I raised money for Balou, but this was of a different order altogether. It was exciting, but also a bit scary. It made me realise how much influence I could potentially now wield. It also dawned on me that I had to be very careful how I used it.

I'd already been approached by people offering us free hotel rooms, boat trips and food, if I gave them my endorsement. A lot were in Greece, but others were further afield. One guy had been in contact offering an eye-watering amount of money to organise a tour of Georgia for me and Nala. He was talking about thousands of pounds in return for us featuring the various destinations and hotels we visited with him on our Instagram.

Georgia was a country I definitely wanted to visit. It was on the road to the Caspian Sea and the Silk Road to Central Asia, which was the route I hoped to follow. But it didn't feel right. What if these places weren't that great, or charged rip-off prices? I'd be obliged to promote something that I didn't believe in. I also didn't want to be constantly shoving products or businesses in people's faces. I wanted to have control over what I endorsed and be genuine when I recommended anything.

Tony had become a good friend by now, and kind of a confidant. I felt as if I could talk to him about most things. When I mentioned my dilemma one night over a beer, he listened sympathetically. A couple of his friends were with us, including a guy from Athens, called Nick. He had been listening intently and was smiling. It was obvious he was itching to speak.

'Do you know the man, Arachnos. What do you call him, Spiderman?' he chuckled.

I looked at him, baffled.

'What are you talking about?'

'Spiderman. The comic books. Peter Parker. What is his famous saying? With great power comes great responsibility.'

I laughed. 'Don't be ridiculous.'

'No, it's true. You are now the Spiderman of Santorini.'

It was only later that I understood his meaning. It was a bit of an exaggeration, obviously. I wasn't a superhero choosing which villains to battle, or innocents to save. But my situation did boil down to the same kind of thing. I had to choose what causes to support, or which opportunities to take, and I had to be responsible about that. I wasn't always going to get things right, but I wanted to be true to myself.

I sent the Georgian guy a note that evening turning him down.

Some causes were no-brainers, of course, especially if they were to do with animal welfare or environmentalism. I'd already started working on ideas to raise money for charities that operated in those areas.

I had spent a few of my days off at a pottery place called Galatea's, outside the little town of Megalochori, a few miles from Akrotiri. I've always fancied learning how to throw vases and other crockery and had a great time under the watchful eye of the owner, Galatea. I made four bowls, which I decorated with Nala's paw-print, and I was rather pleased with them.

With Galatea's encouragement, I decided to hold a raffle, so that my followers on Instagram could get a chance to win the bowls. People could pay £1 each to enter and, at some point, I would do a draw and pick the winners. I'd then distribute the proceeds to charities.

I also wanted to do something meaningful to repay Lucia at Sterila. She'd been incredibly helpful with Nala when I first arrived on the island, and I also appreciated her introducing me to Christina at SAWA.

Sometimes in life, the things you're looking for have a way of finding you.

One morning midway through May, as I set off in the kayak, I noticed a woman taking pictures against the rocks a little distance back up the beach. I couldn't make out what was getting her attention, but she seemed very engaged, as if it was something unusual.

When I came back four hours later, I saw someone else standing in almost the same spot, not taking photos this time, but in animated conversation on his phone. I walked over.

Two tiny little kittens – one black, one ginger – were cowering up against the wall.

'They've been there all day,' the guy said.

I took a look around. They looked so young, there was a chance their mother might be nearby looking after the rest of a litter. However, that seemed unlikely, not least because there was a ginger and a black kitten. I doubted they were from the same mother. I found nothing, so I picked the pair of them up. They were absolutely tiny, even smaller than Nala had been when I found her. I could have carried both of them in the palm of my hand. I took them back to the cave.

Nala's reaction was priceless. She looked at me with utter contempt, as if I'd betrayed her again. When I tried to stroke her, she hissed. She was being territorial, of course, but there was nothing I could do about it. She'd have to get used to the kittens for a little while.

I couldn't look after the kittens for more than a few days, so I called Lucia. She relied on foster parents around the island to provide a temporary home for strays, while she sorted out a long-term adoption. But none of the foster parents had any room at the moment. She still wanted to help, though. Lucia said if I could hold on to the pair for a few days, she was sure she'd find an adopter. There might also be space coming up with a foster parent on the island, a woman called Marianna.

I did my best to hurry things along, posting a few photos with an appeal for a home for the kittens, along with the charity's details. In the meantime, we also took the kittens to the vet in Fira, where they were dewormed and given some flea treatment.

It reminded me of that first visit to the vet back in Montenegro. The vet went through the same drill. He said they were in pretty good condition and were four to five weeks old. The black one was a male, the ginger one female. He needed to record their names, so I came up with two more inspired by *The Lion King*. I called the black one Kovu and the ginger one Kiara.

Back at the cave, they caused mayhem. Kovu was a bundle of energy and had his nose in everything. He reminded me of Nala in many ways. He had no fear whatsoever. Sometimes he liked to lie in a black leather chair in the base's unofficial 'office' area. He blended a little too well into the leather and Tony nearly sat on him one evening. Kiara, by contrast, was a shy little thing and would sleep all day given the chance.

After a couple of days, Nala had completely changed her attitude and now played with the kittens at every opportunity. The three of them would skitter around the cave, chasing after each other, getting under people's feet.

It wasn't long before the charity called me to say my post had done the trick. Apparently, someone from Germany had contacted them directly. They were prepared to go through all the medical and administrative rigmarole that came with transporting a cat halfway across Europe.

'The good news is they want to take the pair of them,' Lucia said. I could sense the hesitancy in her voice.

'And the bad news?'

'Marianna, the foster parent, is willing to take them until they go to Germany. But she needs her new cattery built before she can do so. We don't suppose you know someone who could help her to construct it?'

And so it was – on possibly the hottest day I'd yet encountered on the island, and with Kovu and Kiara in Nala's carrier – that I visited a whitewashed villa on the edge of Fira. The owner, Marianna, was a middle-aged Greek lady. She didn't speak more than a few words of English, but she was clearly the kindest of souls; the house and its small garden was crawling with cats and kittens of various ages. Some were very scrawny, and had clearly been in extremely poor health, but at least now they were in a loving home.

She led me to a small patch of land at the back of the property, where the parts for the new cattery had been laid down on the grass. I took a quick look. It was a metal construction that would have wire mesh walls. Marianna also had some climbing frames, scratching posts and toys to place inside when it was finished. It didn't look like the most daunting piece of engineering, but it would still need some muscle power to put it together.

Fortunately, the charity had made an appeal for extra help and been contacted a few days earlier by a couple from London, who were a little older than me. They helped me assemble the metal frame, and then we wrapped the four sides with the wire mesh. By the end of the afternoon they had to head off, but I managed to fit out the cage inside. No sooner had I put in the finishing touches than the cattery was receiving its first occupants. Kovu and Kiara took to their new home instantly and were soon running around merrily in the late afternoon sun.

Marianna couldn't have been more grateful and offered me a beer and a seat in the shade under a tree. Together we sat there admiring the new construction. We didn't need to exchange many words, with my non-existent Greek and her broken English. We both knew what the other was feeling.

It felt great to have helped her out and even better to know this might provide a home for many an abandoned or mistreated cat. This was the proper way to use my influence. It felt as if I'd got the balance right and found a formula that was going to work.

'Spiderman would be proud of me,' I said, drawing a quizzical look from Marianna. Thank goodness she couldn't understand a word I was saying.

13

Separate Ways

The first rays of sunshine were seeping through the windows when I heard the sound of someone rattling the door and coming into the cave.

'Rise and shine.'

I recognised Tony's voice straight away – along with the sarcasm. We'd had a very late night out in a nearby bar, and he knew that neither of us was going to be shining brightly today. Well, not for a while. I groaned a reply, then hauled myself out of my bed. It took me a moment to adjust to the light, but as I did so, I saw Nala edging her way out towards the half-open door, as was her habit first thing. I would normally follow her out; I'd never grown bored of that first breath of fresh, sea air each morning.

Today I simply wanted to crawl back into bed. So I left Tony to watch her trot off and headed over to the little stove in the far corner. I needed a coffee. A strong one.

It took me about fifteen minutes to revive myself. Then I stepped out – very slowly – on to the beach. The sun was already inching its way higher into the sky to the east. The temperature must have been touching seventy degrees already.

I expected to find Nala in one of her usual spots, in the rocks at the water's edge perhaps. I'd put some food out on the step to the cave, but that hadn't been touched. I went back inside. There was no sign of her in any of her favourite corners further inside the cave either, which was odd. It wasn't like her to miss out on her breakfast.

I stepped outside again. She sometimes sloped off to the foot of the cliff nearby where she lay in the shade, so I took a quick stroll up and down the beach, shouting for her as I went along. There were a few other little nooks and crannies where I'd also seen her play in the past. But she wasn't there. I was puzzled. It wasn't totally unusual for her to disappear like this, but she'd never done so for so long, especially at the start of the day. Maybe someone at one of the restaurants was feeding her. A couple of the waiters there had taken a real shine to her and often gave her little tidbits when I stopped and had a drink or a snack. One would bring out a plate of grilled fish for her every now and again.

I tried hard not to fret too much; she must have found something to entertain herself. She'd be back soon enough.

By now the rest of the team had begun to arrive. To get to the cave, they had walked down from the far end of the beach, where their cars were parked. When I asked if they'd seen Nala at one of the restaurants, they all looked blank and shrugged their shoulders.

Whatever fuzziness I'd been feeling when I'd woken up had long gone. I was wide awake now and trying hard not to panic. I kept telling myself it was okay. It was true that she'd become more independent lately; she'd probably scamper back soon enough when her stomach told her to. I got on with my work.

By nine o'clock I was getting seriously concerned. She'd been gone an hour; this wasn't normal. By now the beach had got busier and busier. People were taking their morning strolls, a couple of swimmers were already in the sea and a guy was walking his large Labrador dog.

I had butterflies in my stomach. I knew something was wrong.

Tony had been off to run a couple of errands. He sensed my unease immediately when he returned, and offered to help. We scoured the beach once more, double-checking a lot of the places I'd already looked. Nothing. There was a large cave at the far end of the beach. Together we climbed in and shone a torch to see if she might be playing in there. Again, nothing.

I'd seen Nala climb up on to the cliffs above the beach a couple of times. On one occasion she'd come tumbling down after being chased by a dog. I was worried that she might have hurt herself, but her self-righting skills had saved her. She'd landed on her side. I marked it down as her using up another of her lives. Two down, seven to go.

We clambered up and looked around, combing through the thick grass in case she was lying in the sun. But all we succeeded in finding was empty bottles and discarded rubbish. Ordinarily I'd have cleared it up, but I wasn't in the mood right now.

My fear was that she had strayed as far as the road, which was set about thirty metres back from the edge of the cliff. It could get really busy, especially in the morning when tourists were being ferried off to the airport for the first flights of the day. My mind started playing tricks on me. I imagined her running out in front of a bus. I told myself to stop. She was a resourceful – and brave – little creature, who had probably survived a lot more than I knew, especially in the Bosnian mountains before I found her.

She'd be fine. She'd look after herself. Wherever she was.

We checked the local bars and restaurants, sticking our heads inside the eating areas before looking in the back and the toilets. Again, she wasn't to be seen. It was the same when we scanned the row of neat villas a little further up the hillside. We asked a couple of the residents whether they'd spotted her. I even pulled out my phone and showed them photos on Instagram. We were met mostly with polite shakes of the head, but one or two people asked us where we were based and said they'd keep an eye out for her.

By now it was around nine-thirty in the morning, an hour and a half since she'd walked out of the hut. She'd not been outdoors on her own that long since we'd been together. I had a sick feeling in the pit of my stomach and I began to prepare myself for the worst. A part of me had always known that this day might come. Bottom line, she wasn't my prisoner. She was a free spirit and, provided she was safe, she was entitled to run off. But it was still a huge shock to think that she'd done it. I'd learn to live with it, I knew, but what I couldn't live with was her coming to harm because I hadn't kept an eye on her. That would break my heart.

The first kayak tour of the day was due to go out within the next fifteen minutes, a little later than usual, around ten. But I knew I couldn't head off onto the water without knowing Nala was safe.

Tony understood completely and told me to carry on searching. He'd send someone else out in my place. He was almost as anxious as me. I could tell he was feeling guilty for letting her out, even though that was perfectly normal.

I walked around for another five minutes or so, more in desperation than hope. I'd looked everywhere I could think of, there wasn't anywhere else she could be. It's strange how the mind works; so easy for paranoia to set in. I found myself walking along the back of the bars and restaurants, looking into their bins. It was crazy, but what if someone had found her dead or injured and dumped her? I found nothing, thank goodness.

I was about to head to the next row of restaurants when something caught my eye. It was nothing more than a flash, in a patch of long grass on an open stretch of land. Tony and I had walked past there earlier, but seen nothing. I walked towards it. I saw another flash, more clearly this time. I could see black as well as traces of brown and grey. I then heard the tell-tale screech of cats. Cats fighting, to be precise.

I started running immediately.

'Nala!'

My heart was thumping and I was breathless by the time I got to the long grass. There were a lot of strays on the island. There was no guarantee it was her.

I don't think it's physically possible to breathe a bigger, deeper or more genuine sigh of relief than the one I let out at the sight that greeted me. Nala was playing with another kitten, a scrawny little black cat roughly her age. The pair of them were tussling and jumping around. They were clearly having a great time and didn't have the slightest interest in anything else in the world. Who knew how they'd teamed up? Or where or when? But they could easily have been playing like this all morning.

I was too happy to give Nala a ticking off. What was I going to say? It was me who'd been lazy and had a lie-in. Besides, the minute she saw me, she turned away from her new pal and came skipping over as if we'd last seen each other only a minute ago. I held her tighter than I think I've ever held anything in my life.

I must have kissed her a hundred times by the time I'd clambered back down on to the beach. When Tony saw us walking towards the shed, he threw his arms in the air as if he'd won an Olympic gold medal. If anything, he looked more relieved than I felt. Poor guy, it wasn't his fault.

'I wouldn't have been able to forgive myself if something had happened to her,' he admitted.

By now the first party of the day was getting ready to go out to sea. I began getting my gear together, but Tony told me to leave it.

'Take the morning off, spend time with her,' he smiled.

I appreciated it.

I gave Nala some food and watched her curl up in a favourite spot on the step overlooking the sea. I sat there with her, watching Tony lead the kayak tour out to sea, my mind in overdrive.

This morning's drama had stirred up something that had been bubbling away for a while. It was now more than three months since I'd arrived on the island and the gloss had begun to come off my idyllic time in Santorini. I was beginning to think it was time for us to move on.

I had several reasons for thinking this way and today summed up one of them.

I'd been having a great time here. It was a party island, especially in summer. I've always loved a party and had made the most of it. This morning made me wonder whether I'd been enjoying it too much. I was too hungover to look out for her properly. What if something had happened to Nala? It would have haunted me for a very long time.

The atmosphere at the kayak school was another factor. It had changed – for the worse. To begin with there had been a high turnover of staff. A few guys came and went, not all of them likeable characters. One of them in particular had been lazy and too full of himself.

Then, a month or so into the season, a couple of members of the team complained about losing bits of money and items here and there. We'd put it down to an occupational hazard. The place was exposed to the public and many people walked up and down that beach. It was going to happen: it wasn't Fort Knox.

But when one day I returned to my little corner of the cave and found my drone had gone, I was really upset.

I spoke to Tony about it and he went to the police, but we heard nothing back. This theft added to the restless itch I'd been feeling.

Of course, there were more positive reasons to get back on the road. For a start, our Instagram following was still exploding. We now had close to five hundred and fifty thousand followers. One of them, a really smart American guy, had visited me at the kayak base and, over a beer, asked me why I didn't have a YouTube channel.

'A guy travelling the world with his kitten,' he said. 'Who wouldn't love watching that?'

He got me thinking. I had a lot of great material on video, but it wasn't anywhere near enough to sustain a channel. I needed new content and to get that I needed to keep moving. I had to make up for lost time, put some serious miles on the clock, and see a lot more countries and cultures. As I thought about it, the fog began to clear a little. I could see a way of turning my good fortune into something of benefit for others.

Christina at SAWA knew that I'd be moving on at some point. She posted something online that touched me deeply: 'We wish Dean and Nala safe travels as they are officially now the best ambassadors for animal welfare in the parts of the world where the animals are forgotten and can truly make a difference to other small shelters struggling around the world.'

The moment I read it, I knew that I had to take advantage of this opportunity.

Last but not least, I didn't think it was fair on Tony for me to carry on at the kayak school. It had got to the point where I couldn't give him one hundred per cent. My main priorities were Nala and the Instagram page. When I was out on the water, he and his colleagues were having to deal with random visitors who'd found their way to the cave in the hope of seeing Nala. It was too much to ask of him. I counted him as a friend now and didn't want to spoil it.

So one evening, after we'd put the kayaks away, I grabbed a couple of beers from the fridge and sat him down.

He knew it was coming, I think, but he was still sad.

I had been due to work until September, but I told him I'd like to leave as soon as possible. He'd had some trouble with staff, as I knew, so I wasn't going to leave him in the lurch. I said I'd stay on until he found a replacement. In the meantime, I started making arrangements to get a ferry back to Athens, from where I'd hop across to Turkey and press on with my plan to cycle across Asia Minor and from there to the Far East.

I contacted the vet and started the process of getting the paper-work I needed. By now the results of the Titer test had returned from Athens. They were fine; the odd cloudiness in Nala's blood turned out to be nothing. The mix-up with the vet in Albania had been sorted too. Nala now had all the correct stamps for her earlier inoculations. All she needed before we headed off was another general examination, which she passed with flying colours. She was good to go.

Tony and I had a couple of nights out before I finally packed up my stuff in the cave and said goodbye to the beach. We promised to stay in touch, and I waved him off almost certain that we'd meet again one day.

On our last night on Santorini, Nala and I accepted an invitation to stay in a little villa on the edge of the caldera, on the posh side of the island, near Oia.

I figured we deserved it.

I'd been on the island for three months and not once had I seen the famous sunset as a tourist. Akrotiri was on the south side of the island and overshadowed by the hills to the east. Even when I'd been with a kayak party, I'd always been too busy taking photos of the customers, or worrying about one of them falling into the sea, to appreciate it properly. The sight of the caldera and the surrounding islands silhouetted in the crimson glow as the sun dipped down below the horizon was every bit as stunning as people said. No wonder Santorini was considered one of the most romantic places on earth, inspiring writers and poets from all over the world.

Sipping my beer, even I found myself acting the Greek philosopher.

Nala had splayed herself out next to me, as content as could be, soaking up the last rays of the sun and without a care in the world. I looked at her and shook my head. In some ways, I envied her. She had no work responsibilities. No bills to pay, no possessions, no pressures. Lucky her. I've always felt that the more people have, the more they have to worry about. Life should be about the simple pleasures. Moments like this. Sunsets, sunrises over deserted beaches. A couple of beers with friends. It doesn't have to be complicated.

My life had become a lot more convoluted these past three months. Too much so. I hoped that being back on the road would untangle things. Allow me to enjoy the simple pleasures in life again. It seemed silly to put into words, but it was true. I wanted to get back to living a bit more like Nala.

14
The Tortoise

As Nala and I stood on the upper deck of the ferry, with the stiff sea breeze on our faces and the port of Cesme on the western coast of Turkey looming into view, I felt an odd mixture of emotions.

On the one hand, I was over the moon. We'd reached an important landmark together; our fifth country and our first new continent. We'd left Europe and entered Asia. But I was also wary of us entering a different culture, a place where people and customs weren't going to be the same as in the West. I needn't have worried. Well, on Nala's behalf, anyhow.

We'd barely got off the ferry and were cycling towards Cesme's old town when a couple of young guys pulled up alongside us on a motor scooter, shouting and waving animatedly. The only word I could make out was 'kedi', which I assumed meant cat. (Was that where the word 'kitty' came from, I wondered?) When I pulled up in slow-moving traffic a few minutes later, a lady in a pale blue hijab came running out from behind a fruit stall. I thought she was going to explode with excitement when Nala let her stroke her. It was as if she had encountered a rock star.

'Looks like that man at the refugee camp was right,' I said to Nala, as I eased my way along the seafront. 'People are going to love you even more over here.'

My feeling that I'd entered a new and unfamiliar world began to grow as I rolled onto the cobbled streets of the old town. The narrow alleys were lined with ornate wooden balconies that were covered in masses of bright purple bougainvillea flowers. The streets were lined with market stalls, the air heavy with the smell of cinnamon, freshly cooked bread and 'shwarma' kebabs being cooked on spits. As the sun set, the distinctive sound of the *muezzins* summoning worshippers to prayers in the mosques filled the air. I found it all intoxicating. Deep down I knew I was going to love travelling through Turkey.

My only concern was the heat.

I'd read that temperatures in July and August could reach 100 degrees Fahrenheit or more, and it felt close to that today, even in the strong wind that blew into the bay of Cesme off the Aegean Sea. It was a dry heat, unlike anything I'd experienced in Greece or elsewhere. It was like being inside an oven, or maybe a fan oven in this case. At times that afternoon and evening, the air was so warm I could feel it on the back of my throat; I hardly dared breathe it in. I needed to walk only a few yards to feel the sweat pouring off me.

Tony had cautioned me to be careful cycling here in the high summer, especially during the middle of the day. He'd seen how the sun could affect me. 'You won't make it out of there alive,' were his precise words. It was good advice, I knew. I didn't want to get heatstroke again.

But the bottom line was that I didn't have much choice. I'd spent months stuck on Santorini. If I was going to make progress across Asia and beyond, I had to start racking up some miles. With this in mind, I'd made a plan to cycle from Cesme across to Izmir, down the so-called Turkish Riviera to Marmaris, and then on to Fethiye and Kas. From there, I hoped to head north via Cappadocia to the Black Sea, ready to carve a route east into Georgia and Azerbaijan, gateway to the Silk Road and Central Asia.

It was an ambitious undertaking. The route I wanted to take was at least 1,200 miles, or 2,000 kilometres, to the border with Georgia. It was going to be a long slog, but I was determined to give it a go. As I settled into my first evening in Turkey, I knew that I would have to cope with the weather – one way or the other.

It was now a couple of weeks since we'd left Santorini. We'd travelled back to say hello to Iliana, Nick and Lydia, but then accepted four days on a boat cruising around the Saronic Gulf, the bay of islands off the coast of Athens. The owner of the company, a kind, charismatic Greek guy named 'Captain' George, was one of the first people to contact me when we went viral. I liked the fact that he'd been very laid back about it and there was no pushy salesmanship. So I thought it would be a good chance to recharge our batteries in readiness for Turkey and the next leg of the trip. As it turned out, George's yacht also gave us a chance to recover from the shock we'd both had in Athens.

I'd been sitting with Nala outside a cafe when she suddenly bolted up and jumped onto my shoulder, agitated. A few seconds later, I heard the sound of clinking glasses and tables and chairs being dragged across the floor. The walls were soon vibrating in the same jelly-like way I'd seen in Albania. It was another earthquake. It lasted for no more than ten to fifteen seconds, but felt much more powerful than the one I'd experienced before, and it had freaked Nala out even more. When we returned to Athens a week later, after our cruise with George, the city was still clearing up the mess.

We sailed out of Piraeus again, but this time on a ferry heading east to the Greek island of Chios, where I was told I could get a connecting ferry across to Cesme and Turkey. The trip didn't go as smoothly as I'd hoped. I'd not spotted that Chios has two ports – at opposite ends of the island. So I had to cycle thirty miles from the one linking to Greece on the west side to the Turkish service on the east side. As if that wasn't bad enough, we'd arrived in the evening and I had to ride there on unlit roads in the pitch black. The lights had also gone on my bike, and it was not the most relaxing ride.

Fortunately, there was little traffic on the road and we arrived there safely, but we then had an eight-hour wait for the small, elderly-looking ferry to take us on the half-hour trip to Turkey. After my last experience at passport control in Albania, I was quietly dreading Turkish immigration. We were leaving Europe and entering western Asia, so I expected checks to be more stringent now. I also remembered the man I spoke to at the refugee camp telling me how difficult it had been getting in from Syria.

Of course, there was no comparison between his situation and ours. We arrived in a tourist region of the country and were heading in a different direction entirely. So perhaps I shouldn't have been surprised that the customs officials seemed only interested in passing the buck.

No one seemed to have a clue what to do with Nala's passport. I'd present it at one window then be told to go to another one. We were bounced around for about half an hour before an official threw his hands up and waved us through. I was beginning to think I was the first person in history to have gone through some of these customs points with a cat. They certainly acted that way.

I wanted one night to acclimatise, so I booked us into a hostel in Cesme. Nala spent her first hour jumping around the bunk beds, playing hide and seek with me, but she soon dropped off to sleep. I had no such luck. The heat remained oppressive well into the small hours and no matter how much I tossed and turned, I couldn't doze for more than a few minutes at a time. I wasn't disappointed when Nala started meowing for her breakfast around six o'clock as usual. I wanted to be on the road early.

About half an hour later, I was bumping and bobbling my way along the cobbled streets of Cesme. It felt good to be cycling in the relative cool of the morning. After three months holed up in Santorini, I was glad to be back in the rhythm of life on the road – though it had its challenges, of course.

I'd got used to travelling with all my possessions on board my bike, but as I began this new phase of the journey I began to feel more like a tortoise, carrying our home around with us. That home seemed bigger and more cumbersome than ever. I shouldn't have been surprised. According to the manufacturer's info, I could carry an additional 25kgs on the back and almost 12kgs more on the front of the bike. I reckoned I was now close to that limit, so that on top of the bike's basic unloaded weight of 13kgs, I was now hauling 50kgs, or around 110lbs in total. That was without taking into account the extra few kilos of clothes and other bits and pieces I had packed into the pouches on my trailer.

I'd done what I could to lighten the load, discarding anything unnecessary, but my problem now was that I had a lot of stuff that I couldn't do without.

The most important addition since we'd last been on the move was Nala's new accommodation at the front of the bike. I improvised it by taking a standard, pet shop stroller-cum-carrier and fitting it to the bracket that had held the old pouch. It made for a more spacious and comfortable carrier, which also gave Nala some shade if she needed it. She took to it immediately.

With the money I saved up from working at the kayak school, I bought myself some new gear too, including a better laptop, a new GoPro, and a drone to replace the one that had been stolen. If I was going ahead with my plan to launch a YouTube channel, I needed some decent quality equipment.

Packing up all this gear had become a military-style operation. It had to be stowed away neatly along with my camping gear, sound system, batteries and chargers, tent, stove and food. Then there was my clothes and Nala's ever-increasing collection of stuff to be stored. Loading the bike was also a bit of a science. With so much weight in each of my bags, I had to make sure I distributed it evenly, otherwise I'd be off balance all the time. The first few yards of my journey were a real effort. It wasn't easy to build up some momentum when I pushed off, especially if I was heading uphill.

None of this took away from the exhilarating feeling I had being on the open road. I loved it as much as ever, especially with Nala on board. As we began the fifty-mile journey east from Cesme towards the ancient city of Izmir, she was her usual, animated self, reacting to everything around her. Every now and again, she'd lunge at a butterfly or a hornet whizzing past us.

The road to Izmir weaved along the coast and inland, too. The further from the sea we went, the more ferocious the heat became. By midday there was absolutely no way I could be out in the sun, and it was only around five o'clock in the afternoon that I could venture out again. Even so, every cloud – or should that be lack of cloud – has a silver lining. The unrelenting heat gave me a chance to do more sleeping outdoors.

I've always loved camping out under the stars and especially in odd places. Even as a child, I used to sneak out of the house and sleep out overnight, making myself a bed under the picnic bench in our garden. It was a habit I continued as I got older, much to the bewilderment of my friends. One told me that I had 'bad wiring', and there was something wrong with me for sleeping out in derelict buildings or on isolated beaches. I took it as a compliment. If people didn't get the buzz that I did from being close to nature, to feeling, seeing and smelling the real world around me, or of being at the mercy of the weather and its ever-changing moods, then that was their loss. To me, there was nothing better.

So I was delighted on the first night out of Cesme when I found an abandoned swimming pool a hundred yards off the main road near a little beach. I lowered the bike and all my gear inside the empty pool and set up my sleeping bag and pillow on the floor. For me it was – almost – as good as a five-star hotel. I slept like a log.

The Turkish roads were in good condition and often very wide, which made cycling more relaxing. As long as I hugged the dedicated, inside lane, I didn't feel threatened by the heavy lorries that swept past at regular intervals. By the end of the next day, I'd made good progress and had arrived in Izmir.

I booked us into a nice hostel for a night; it gave me some time to explore the ancient city and its sights. Nala was even more fascinated by the place than me and kept darting off in one direction or another as we walked through the old marketplaces and squares. In the small, enclosed streets of a city, with no breeze, the heat was absolutely draining. By the end of the day she was as exhausted as me.

The pair of us overslept the following morning, which was unusual. It must have been the heat. I made a mad dash to pack everything up, then hurriedly checked out of the hostel and hit the road by mid-morning, frustrated to have missed the best part of the day for cycling.

I'd been hoping to travel sixty miles-plus today and had booked a bed for the night in a hostel in the town of Aydin. But my late start meant that by midday I'd ridden only about twenty miles. It was, if anything, hotter than ever, so shortly after noon we took shelter under a road bridge and only managed to get going again late in the afternoon. The prospect of making the hostel by dark was looking doubtful. I pushed myself hard, but by dusk I realised I'd have to sleep outdoors again. It wasn't the end of the world. I cancelled the hostel and started looking for a spot for the night.

The sun had just set as we passed an abandoned building site of some kind. It looked ideal. I fixed up my hammock and sorted some supper for the pair of us. The usual pasta for me, some chicken chunks for Nala. While Nala snoozed afterwards, I chatted on the phone with my family back in Scotland and updated people on Instagram.

A lot of followers fretted about us sleeping outdoors, but I'd decided to ignore them. If I read and listened to every opinion I read on our page, I'd get precisely nowhere. I had to stay true to myself and follow my instincts as much as possible.

I used an app on my phone that tracked my total mileage. As I lay under the Turkish stars that night, I noticed that I'd done 4,000 miles since I left Dunbar. It felt like a real milestone, in every sense of the word, but I also knew I'd barely scratched the surface of my journey. Even getting out of Turkey was a distant prospect: that was still 1,000 more miles away. I had a long, long road ahead of me. I had to get a move on.

The next morning, I cooked up a little porridge for breakfast and started going through the laborious process of loading up the bike. As I strapped on the two big bags of kit that sat at the back, I noticed the rear tyre looked soft. It was hardly surprising; it was carrying a lot of weight now. I gave it a squeeze and decided I'd better give it some extra air. It was annoying; I'd have to take the bags off again, delaying me even more. But it had to be done.

I got going a little before eight, leaning hard into the pedals as I pushed off down the road. Even at this early hour, the heat was stifling. Nala decided wisely to stay in the shade inside her carrier. My vest was soon bathed in sweat and only ten miles down the road I had to make my first drinks stop.

As if the bike didn't weigh enough, I was now carrying several litres of water with me. I wasn't sure I could trust the water in the rivers, or in the taps I found in toilets at petrol stations or cafes, and running out of water didn't bear thinking about. I guzzled down half a bottle then poured some into my cupped hands so that Nala could lick away. She was lapping up the last drops from my palms when I glanced at the back tyre. I thought I'd better give it a kick, to make sure it was holding up.

I was about to do so when I froze. I couldn't believe it. One of my bags was missing – and not just any of the bags, but the one with all my electrical equipment in it, including my new drone and computer.

I groaned. I must have left it back at the building site.

I wasn't someone to panic too much, but I set off back down the road as fast as I could, my mind rewinding through the events of earlier. I could picture what I'd done quite clearly. When I took the bags off to inflate the tyre, I put them on the other side of a low wall. I must have been distracted by Nala and picked up only one of them again. How, I had no idea, but more baffling was how I'd not spotted it. My bike must have felt lighter when I hit the road. It would also have been listing slightly to one side without the other bag's weight to balance things.

Why hadn't I noticed it, and what was I thinking about that was more important? Had I been in too much of a rush?

I was still berating myself when, about two-thirds of the way down the road, I was in even more trouble with a puncture. I'd noticed already that the sides of the roads in Turkey were littered with tiny metallic fragments. I asked about them at a garage and was told that there were always a lot of blowouts, because people drove their tyres to the limit. The metal scattered around was from their inside wiring. I'd been afraid it was a recipe for punctures and been proven right.

I pulled up on the side of the road – then swore loudly. The puncture kit was also in the bag I'd left behind. I couldn't fix it here. I had no option but to press on, wheeling the bike the final couple of miles. I couldn't afford to destroy the tyre totally by putting my weight on it.

I was a dripping mess by the time the building site loomed into view. My heart was pounding so hard I thought it would jump out of my chest. What if the construction workers had come back? Or some kids had begun playing on the site? I heaved a huge sigh of relief when I saw the bag exactly where I'd left it.

I spent some time calming my nerves, then stripped everything on the bike back down so that I could fix the puncture. With all my bags laid out and the tyre finally fixed, I decided to go through a checklist of stuff. Something else was nagging me. I couldn't put my finger on it.

I'd laid everything out when I realised. Our passports.

In a panic, I grabbed the small pouch where I keep my most important documents. I knew almost the instant I opened it. The passports weren't there.

My heart jumped again. At this rate I was going to end up connected to wires in a coronary ward. I fell to my knees. I couldn't believe it. Had I had a complete meltdown? How had I managed to screw things up so spectacularly?

Nala had been playing on a wall nearby, but even she picked up on my agitated state. She came over and meowed as if in sympathy.

'Sorry,' I said, giving the back of her head a ruffle. 'Dad's having a bad day.'

I took a deep breath. I tried to stay calm and recreate my journey since I last saw the passports. There had to be a logical explanation, but I was so upset, I couldn't summon up anything. It was as if my brain had flooded like an overworked car engine. Hard as I tried, I couldn't remember. Maybe they'd fallen at the side of the road? Perhaps they'd been stolen? I looked around the area where we'd slept the night before, but there was nothing.

The next hour was one of the most fraught of my trip so far.

At times like this your mind plays tricks on you. It comes up with all sorts of catastrophic scenarios. For a while, I wondered whether the Home Office would even issue me a second new passport within a year. They might think I was an idiot who couldn't be trusted with something so important. I began to think about Nala and what would happen to her. Where would she stay if I went back briefly to the UK? And what would happen if I couldn't get back into Turkey?

For the first time I felt almost defeated. I began to wonder if this was the end of our adventure, and if maybe that would be for the best. I'd made such slow progress in the past few months; how was I ever going to get around the world anyway? I'd be sixty by the time I hauled my overloaded bike back to Dunbar.

Gradually I managed to work myself out of my funk. It was a problem, but it wasn't the end of the world. I'd sort it out somehow. I'd done it before and I'd do it again. I googled the nearest hotel and booked a room via their website. I didn't care what it was like or how much it cost, I needed to take stock. To work out what to do next.

My guardian angel must have been floating around again that morning. Half an hour or so later, and not before I'd suffered yet another puncture, I wheeled the bike into the courtyard of a little guest house a half mile off the main road. It was a rural place, set on a hillside in an olive grove. The timber-framed buildings were simple, but there was a swimming pool and an outdoor eating area.

The place seemed empty. The lady at the reception looked Turkish, but greeted us warmly in English, introducing herself as Sirem. She made a beeline for Nala, whom she started stroking and cooing over. She was soon joined by a middle-aged guy with a healthy suntan and long hair. He spoke English too, but with an accent. It turned out he was Australian. 'My name's Jason,' he said.

It was standard procedure to hand over your passport when booking into a hotel, so I explained my situation. I must have looked agitated, because Sirem immediately told me to take a seat then offered me a cup of Turkish tea.

'It's okay, you can stay here tonight. Don't worry,' she said.

'Where did you stay last night?' Jason asked.

'In a disused building site,' I replied, drawing an odd look from him.

'And the night before that?'

I dug out the bill that the lady had given me when I checked out from the hostel in Izmir.

'Here,' I said, offering it to Sirem.

'They will have your passport number. I will give them a call. Don't worry,' Sirem said, reassuring me again. She then disappeared into their office.

I got talking to Jason. He told me they'd not long opened the place. They'd built it up from scratch over a period of a couple of years, using sustainable materials – wood for the frames and straw bales coated in plaster for the walls.

They were growing their own vegetables and olives, making their own bread. They were trying to make the place as self-sufficient as possible. As if that wasn't impressive enough, they also seemed to be animal lovers. There were photos of cats every-where, and Jason told me they'd adopted several strays and a few dogs too.

I felt better already. If I did need to dash back to the UK to get a new passport, I couldn't have chosen a better place for Nala. There was something about their warm manner; I trusted Jason and Sirem instantly.

I unloaded my stuff into my room and spent some time playing with Nala. She was always there to lift my worries. I put the video camera on as I chased her around the bed in a game of hide and seek, ducking under the bed frame, then popping back up again to tap her on the nose or the back of the head. Every time I appeared, she'd go into a frenzy, flailing at me and trying to bite my hand while her tail wagged furiously. It cheered me up no end.

I'd started writing a text explaining my situation to my sister back in Scotland when Sirem appeared. She had a big grin on her face.

'I have got some good news,' she said. 'Your passports. They are back at the last hostel. Back in Izmir. Apparently, you left them at their reception when you were checking out. They are keeping them for you.'

I felt like a fool. How could I not have thought of that?

'Reception. Of course,' I said.

I saw it all as if it was a scene from a movie. Two mornings ago, when I was checking out; it was all slightly chaotic. I was flustered and frustrated after oversleeping. I put Nala and the bike outside with my bags attached, then popped back in to settle my bill. The receptionists were busy and I had to wait while they dealt with some other guests.

But then I got worried when a small crowd gathered around Nala. I had to keep dashing in and out of the hotel lobby to check on her and the bike, which was carrying all my kit. So when I finally paid my bill, I was in such a hurry that I pocketed the bill, but not our passports. No one at the reception spotted them until after we'd gone.

The relief I felt was immense. My only problem now was getting back to Izmir. It was a day's ride away and I didn't think I could make it today. It was impossible to cycle in the afternoon heat. If I jumped in a cab, it would probably take three or four hours in total. Goodness knows what it would cost me, but that wasn't my main concern.

Jason had appeared again. I asked him and Sirem for the number of a cab firm, but he shook his head.

I was surprised.

'There aren't any cabs?'

'Yeah, there are plenty,' he said. 'But you don't need one. I'll drive you up there.'

I was dumbstruck. I couldn't believe someone would do that. I've heard of random acts of kindness, but this was ridiculous.

'At least let me pay for the petrol.'

'No need,' he said. 'I can run a couple of errands.'

Before I knew it, we were heading back up the main road towards Izmir. Nala sat beside me as we whizzed through the Turkish countryside, passing places that I'd cycled past only the previous day, including the building site.

When we got to the hostel, the lady had the passports waiting for me. I couldn't thank her enough for keeping them. We were soon back on the road, heading south.

As if Jason and Sirem hadn't been kind enough already, when we got back early that evening, she cooked us a lovely traditional Turkish meal. She and Jason also offered me an extra couple of nights for free. I accepted. I'd already decided that I was going to give them a glowing review on Instagram. Regardless of their kindness, their place deserved it.

We sat out in the garden of the hotel late into the evening, drinking, chatting and exchanging stories. I couldn't stop apologising, no matter how often Jason and Sirem told me not to.

'I'm terrified I'll do a *Home Alone* one day and leave Nala behind,' I joked at one point.

'I doubt that,' Sirem said, nodding at Nala, who was curled up alongside us both. 'I think you would leave your bike behind before you left her.'

I must have been feeling better, because my mind had already turned to the road ahead. They were both well-travelled and seemed to know Turkey inside out. With a big map in front of us, they recommended a couple of places in and around Cappadocia, as well as up towards the Black Sea coast on the road to Georgia. They also repeated the warning I'd had from several people now, not to stray to the south-eastern end of the country and the border with Syria.

'So, what's your plan after you reach the Caspian Sea?' Jason asked me.

'The Pamir Highway, then down into India and Thailand. By the end of next summer, I hope.'

Jason leaned back and whistled.

'That's ambitious.'

'Yeah, probably. Especially as I'm like a tortoise these days, carrying my entire home on my back.'

Sirem had been sitting quietly, playing with Nala. They'd taken a shine to each other.

'Do you know the fable?' she said, smiling. 'About the tortoise and the hare?'

I remembered it vaguely from school.

'Aye. Slow but sure wins the race.'

'Exactly,' Sirem said. 'The thing is: you aren't even in a race.

So what is your hurry? Why not just enjoy the journey? Take things steadily and see where fate takes you.'

She nodded towards the beautiful sunset and the rolling valley beneath us.

'Just look at where it brought you today. I'd say that sometimes there are benefits to being a tortoise.'

They were the wisest words anyone had said to me in a while. And exactly the ones I needed to hear.

15
Into the Wild

For the first time in days, I felt that I had made a good decision. Accepting Sirem and Jason's offer to stay on for an extra couple of nights gave me time to fit the bike with a new piece of equipment. I'd found the miniature, adjustable umbrella by chance in a shop selling baby supplies, when Sirem took me to their nearest town, Germencik. It was UV-resistant and fitted neatly above Nala's basket, leaving her free to move around and look out while keeping cool at the same time.

It was soon doing its job.

The day we left Sirem and Jason and headed south through the mountains, the mercury touched 100 degrees at one point. At times it felt as if I was cycling inside a blast furnace. I wondered if I might melt, but I knew Nala was now safe under the shade of her new parasol.

Sirem's words of wisdom stayed with me. Slow but sure, I told myself, it wasn't only about getting through the miles. With that in mind, I accepted an offer via Instagram to spend a couple of days at a resort on the way down through Marmaris. A little R&R wouldn't do us any harm. We rolled in there a few days after leaving Jason and Sirem.

The owners of the resort had rolled out the red carpet and put us in a luxurious double room in a chalet set in ornate gardens. When we checked in, there were treats laid out on the bed – some snacks for Nala and a few cans of Scotland's favourite soft drink, IrnBru, for me. There was even a balloon decorated with the globe and our Instagram 'handle' *1bike1world*.

'We must be important,' I joked to Nala, as we explored the resort's gardens, spa and beach.

It was a toss-up as to who enjoyed the stay the most. In between sessions of swimming and relaxing in the spa, I treated myself to a much-needed massage. The toll the road had taken on my legs was terrible; my calves and thighs felt like slabs of concrete when I lay on the table. The masseuse had to work extra hard to breathe some life back into them. Nala, by contrast, had a whale of a time walking the gardens and climbing the lush, leafy trees. A small group of cats roamed the estate, and she hit it off with a little white-and-grey kitten. They spent ages tussling and chasing each other in the grass outside our suite.

By the time we were ready to leave, I felt rejuvenated and ready for another spell on the road. I was in good spirits as I loaded the bike and got Nala ready. But I soon had the wind taken out of my sails.

I'd learned by now to keep an eagle eye on Nala's health. She'd been through so much early on. I felt that she'd had enough of vets and vaccinations and I'd been particularly watchful since we arrived here in Turkey. Intense heat can be dangerous to a cat. Dehydration can lead to kidney failure and all sorts of other problems. So when I ruffled or stroked her, I'd often try to sneak in a quick check of her skin for cuts, bumps, bruises and bald patches. Every now and again when she was asleep, I'd look at her teeth and gums for any signs of infection. It was an easy way of monitoring her.

As I placed her in her basket this morning, I noticed a nasty-looking sore on her upper lip. It looked like she'd had a cut and a scab had formed on it. It seemed as if it might be painful too. When I went near it, Nala flinched and hissed at me, which I took as a bad sign.

'Better get that looked at, Dean,' I said to myself.

I wasn't too despondent. As luck would have it, I was heading to see someone who I hoped would be able to do precisely that. One of the managers at the resort had recommended that I visit the animal sanctuary in the nearby town of Gokova. She told me that the owner, a Scottish lady called Jeanie, was a bit of a heroine in the animal welfare world.

It took me a couple of hours to travel the twenty-five miles to the sanctuary. The journey was almost entirely uphill, and in the searing heat I thought I'd never make it.

When we finally arrived, our welcome couldn't have been more friendly. Jeanie greeted us like members of her family. She was a colourful character, that was for sure. She had been born in Dumfries, but she'd been out here in Turkey for more than thirty years, during which time she'd built up her sanctuary from nothing. After I'd cooled down with a drink, she gave me a tour. Nala came with me, fixed to my shoulder, curious as ever about what was going on.

The sanctuary was spread out over an acre or more and over-looked the bay of Marmaris in the distance. Jeanie told me that she'd begun with a single stray cat. Now she had one hundred and thirty of them, along with half a dozen dogs, a pair of donkeys and a horse.

She was particularly fond of one of her donkeys, an elderly male – or jack – called Ned. He was grazing quietly in a little pasture on a slope.

'Found him abandoned in the middle of a field, about five miles from here,' she said, as I gave him a stroke. 'He'd been left to die, tied to a tree with a shard of metal sticking in one of his feet. Poor fellow.'

It was stray cats that had been the main focus of Jeanie's work, though. I've never seen so many of them in one place. She had every breed, colour and temperament known to cat kind, from big Persian fluff balls and streetwise-looking tabbies to skinny, almost hairless kittens that were too shy to even acknowledge her. Nala was clearly unsure what to make of them all. She sat there, digging her claws extra deep into my shoulders and making quiet, almost pleading meowing sounds in my ear. *Please, Dad, don't even think of leaving me here.*

With the midday sun now burning away in the sky, Jeanie took us inside for some lunch. Afterwards I asked her to take a look at Nala's lip.

Nala was like putty in her hands and let her look at her mouth without any protest.

'Hmmm. Don't like the look of that,' Jeanie said. 'Might be what they call rodent's mouth. Quite common here. I've got an errand to run for an hour or so, but I've got a friend who can drive you down to the vet in town later. I'll give her a call.'

I was soon wishing I'd never asked.

Jeanie headed off, leaving me to wait for her friend to take me to the vet.

What's the expression? 'The devil makes work for idle hands.' Scrolling through vet sites online, I saw that the technical name for rodent's mouth was *feline eosinophilic granuloma complex.* No wonder they'd simplified it. Normally it wasn't serious; it would disappear on its own without treatment. But there were cases where the sores had developed into carcinomas, cancers, especially when they were caused by stress.

My mind was running away with itself again and heading straight to the worst possible scenario. For a while I racked my brain for what might have stressed Nala. It wasn't difficult; there were plenty of possibilities. The earthquake. That fall she had back on Santorini. The various thunderstorms we'd cycled through. The list went on. I was on tenterhooks when I heard a car pulling up on the gravel outside.

Jeanie's friend was a local and spoke only a little English, but she was lovely and took me straight down the hill into the village.

The vet was a nice young woman, no older than me, and spoke good English. She put my mind at rest immediately.

'It's probably just a cut,' she said. 'It should heal itself but keep an eye on it. If it's still there in, say, two weeks, take her back to a vet.'

It was good news, but I was still blaming myself for it. How had she got a cut? Could it be from bouncing her up and down as I rode on the cobbles back in Cesme? Or biting and licking her lips too much in this heat? But then I remembered something else.

During our stay at the resort, Nala spent a lot of her time play-fighting with her friend, the white-and-grey kitten. One day I heard Nala let out a loud yelp. She came haring in afterwards, as if she'd been hurt. It seemed the most likely explanation.

Jeanie was pleased that I'd been to see the vet and even more pleased by the diagnosis. That evening she prepared a lovely dinner. We chatted on her terrace; she was full of stories about her early days and how she'd built the place up. I was full of admiration for what she'd achieved.

There was a picture on the wall of one particular cat, which seemed to have pride of place.

'Looks like a character,' I said, nodding at the photo.

She smiled.

'Her name was Korkiz, which translates as blind girl. A young Dutch tourist called Iris discovered her in the foundations under-neath a hotel nearby. It had been the last day of her holiday and she'd not known what to do. So someone at the hotel had called me and I'd taken her in. She was in a bad way, already blind in both eyes, which is why she got the name.'

Nala had been having her own dinner in another corner of the house, but now jumped up and slid into a chair alongside us.

'I didn't think any more of it, but about a year later I got a cheque – a big one – from Holland,' Jeanie went on. 'It turned out that Iris had spent the year raising money for my sanctuary. It was a turning point for me. I'd just moved from our original place to here. I was desperately short of money. If it hadn't been for her, well, who knows?'

After we'd cleared away dinner, Jeanie showed me photos of the various awards she'd received, not only in Turkey, but also back in the UK. In one of them she was standing with Iris and the local mayor. In another she was standing at a ceremony inside the House of Lords, in London.

She nodded at the photo of Korkiz again.

'Sometimes a single animal can change everything,' she smiled.

She leaned down and gave Nala a stroke on the back of her head. 'And that's why you've got such a treasure on your hands with this one.'

'Well, I'm pretty fond of her, yes,' I smiled.

She shook her head gently.

'No, I don't think you get how special she is. It's so difficult to raise money to protect animals these days. People get so many charities asking for help. They just tune out. But every now and again a story connects with people, like yours did,' she said. 'I saw what you did in Santorini. I've got a feeling you can do a lot more.'

She then gave me the kind of look my mum might have given me when I got a new school blazer.

'So make sure you look after her.'

I left Jeanie the following morning, promising to stay in touch. I'd already made a quiet pledge to myself to help raise money for her if I could. She would always appreciate financial help. Organisations like hers live from hand to mouth. There are too many stray and abandoned animals in the world to ever stop helping.

For now, though, my priority was to get some miles under my belt. Over the next couple of days, I headed south and then east from Marmaris, following the Turkish coast past the town of Kas and then on to Antalya. I wanted to get to Cappadocia by the end of the week, so I pressed on, leaving the coast and heading up into the mountainous interior. The terrain felt different almost immediately; it was a landscape of rugged mountains and dense forests. I could cycle for miles without seeing a single building. It felt like I'd stepped into the wild.

Progress was slow, but I kept repeating my new mantra: I was the tortoise not the hare.

A couple of nights into the journey, I decided to pitch my tent in some woods, a couple of hundred yards off the main road. It was a scenic spot; I had a stunning view out over the valley below and enjoyed the sunset. As I hunkered down inside the tent with Nala, I heard the odd noise, but nothing that worried me too much. It sounded like small animals of some kind; I knew the woods were alive with rabbits and foxes. I zipped up the tent and settled down for the night.

It was hot even here in the forest and finding it hard to nod off, I put on my headphones and started watching videos on YouTube. I was getting ready to launch my own channel in the next day or two and wanted to see what else was out there. At around one in the morning, Nala suddenly jumped at me, clambering on to my shoulder. She'd been sleeping soundly at my feet and hadn't moved in ages, so it gave me a real start. Something had spooked her. Her ears were pricked and her eyes as wide as saucers.

It didn't take long to work out what had panicked her. With my headphones now off, I could hear a deep, powerful, huffing sound. It sounded as if something large was moving around slowly nearby. It was pitch dark and there was next to no moonlight, but I could almost feel it looming over us. It was definitely a big animal of some kind.

My mind went into overdrive. What was it? A jackal, a wolf? No, too small. A deer or a cow of some kind? No, I knew their noises. The reality hit me like a punch to the stomach. The only thing it could be was a bear. I'm not easily spooked, but I completely freaked out.

I grabbed Nala and threw open the side of the tent. I'd been sleeping naked, but didn't even think about clothes. I slipped into my Crocs and ran. It was only when we reached the main road that I realised I'd spiked my foot with a nail while running through the woods. I was bleeding and a bit shaken up. Nala was also agitated and clinging to me as if there was no tomorrow.

It took me a while to calm down. But then I realised that I was standing in the middle of a road stark naked. I would have to go back to get all our stuff. Or what was left of it. If there really was a bear in our neighbourhood, I dreaded to think what it could do to my bike.

After a few minutes, I summoned up the courage to pick my way back into the forest. Every rustle of a leaf and crunch of a twig made me jump.

The shadows seemed to be alive. I'll admit it, I was scared stiff.

Back at the tent I threw on a T-shirt and a pair of shorts and started gathering my gear as fast as I could. My batteries had run flat again, so I had no lights on the bike. I grabbed my small torch and flailed around in the darkness, moving quickly. I put Nala on the front of the bike, then packed up the tent. I was on autopilot and in a matter of seconds, I was pushing the bike and all the bags through the woods towards safety.

Arriving back on the road intact, I let out a roar of relief.

I was certain I must have left stuff behind. If I forgot important things in daylight when I was calm, what on earth was I going to forget in the middle of the night when I was scared of a bear? But *c'est la vie*. Right now, I couldn't afford to worry. I knew bears were good at tracking. If it had picked up on our scent, it might follow us. Realistically, the chances of it attacking us on a main road were small. But that didn't stop my mind from going there.

I cycled up the hill in front of us as fast as I could, the cut in my foot hurting like crazy. My heart was still pumping and I kept looking behind me, half-expecting to see a big grizzly lumbering up the road, on my tail.

Reaching the brow of a hill, there was a dim light glowing ahead of me. I could have cried tears of joy when I saw a big construction site ahead. It was some kind of civil engineering project, where they were laying an enormous pipeline. Gigantic metal pipes, each one at least two metres tall, were stacked up, five high.

If I could climb to the top of the stack, I'd be more than ten metres in the air.

There's no way a bear will get up there.

I clambered up with my mat, sleeping bag and Nala across my shoulders. Before long, we'd laid ourselves out inside a pipe at the top of the stack. All I could do was lie there, listening to the sound of my heart, settling slowly into a gentler rhythm. There was a decent phone signal, so I had a quick exchange on our lucky escape with my family back home. Their reaction was not what I expected; my mother started laughing. She said a couple of nights back my dad had a dream that I was being chased by a bear.

'He must be psychic,' she giggled.

I had to see the funny side. There was no actual proof that it had been a bear, though I knew they lived in certain parts of Turkey, but I was glad I hadn't hung around to find out.

I managed to get a decent few hours' sleep, which was just as well, because I was back on the road early again in the cool of the next day. I forced myself to put last night's drama to the back of my mind; I had another tough day ahead of me.

To get to Cappadocia, I had to continue through this mountainous section of country. From a cycling point of view, the next bit was probably going to be the biggest test I'd yet faced. The route I'd chosen took me through a national park, the Koprulu Canyon, and then up and over a 5,000-foot peak. My aim was to do it in a single day.

I started well. The cut on my foot wasn't as bad as I thought and had already started healing. I was cycling freely. I stopped for lunch at a beautiful spot next to some rapids in the river. I filled up Nala's little water bottle, then smeared both of us with an entire tube of sun cream. We'd be going up and over the mountain during the hottest part of the day.

As the peak loomed into view, the terrain changed dramatically. The smooth tarmac road became an incredibly steep gravel and shale track. I was able to cycle up tough gradients, up to an eight-degree incline, but this was ten degrees. It was too much. I was forced to get off and push. I'd loaded myself up with extra supplies of water again, so the bike seemed heavier and more awkward to manoeuvre than ever. Every now and again I'd slide backwards on the loose shale and gravel surface. I was relatively strong, but even I struggled to keep it from tumbling down the mountain a couple of times.

One or two cars passed me by. I could see they were struggling too; their wheels were spinning, especially when they went round the hairpin corners. The heat was close to unbearable. No sunscreen in the world would have been able to protect you from it and I could feel my shoulders and neck burning up.

To make things even worse, I got chased by a swarm of bees when I took a break next to a mountain stream. I managed to escape them without being stung, but then a little further up the mountain I got hassled by a pair of nasty-looking wild dogs. I have no idea what kind they were; they looked like they might be related to hyenas. Luckily, they lost interest in me when they spotted a dead hare lying on the side of the road. Nala, thankfully, slept through the whole drama.

I'd encountered some killer hills on the trip already in Switzerland, Bosnia, Albania and Greece. But nothing to compare to this one. By mid-afternoon I was ready to throw in the towel.

Some young guys came past in a car, waving at me as they went. I stuck out my thumb, but they drove on. They didn't have room in the car, especially given the size of my bike with all its para-phernalia. But then a family in a battered old flatbed truck appeared. They were struggling to get up the road, too; the truck's engine was whining as if it was in pain. I stuck my thumb out again.

There was a mother, father and what looked like their teenage son and daughter in the cab. The woman acknowledged me, but shook her head apologetically. I could probably have ridden in the back. It was covered by a tarpaulin roof. But I don't think they dared stop for fear of never getting going again. Their back tyres were spinning on the loose, shale road.

I kept looking at the map on my phone to see if the peak was drawing near. But it was like treading water; the little dot on my phone seemed to stay still.

By late afternoon things had gone from bad to worse. First, I got a puncture. I'd sensed the front tyre was going down when I came across the wild dogs. It would have been impossible to change a tyre then, but I couldn't avoid it any longer. It was as flat as a pancake.

I pulled into the side of the track and stripped down the bike. Another car went past. For a moment it looked like the couple in it were going to take pity on me and pull up. But again they thought better of it and drove on.

I was finishing off the tyre when I felt the faintest breeze and a sudden drop in the temperature. It wasn't as welcome as it might have been. In the valley ahead of me, a bank of brutal, black clouds was drifting our way. I could see it was already spitting out forks of lightning: this was a thunderstorm. I knew I was in trouble. The hillside was barren apart from the odd tree, and a lot of them had a strange, burnt-out look, as if they'd been hit by lightning strikes in the past. We were really exposed.

Is there anything else that can go wrong today?

I made my usual mistake of googling lightning strikes and their impact. I learned something, but it didn't make me feel any better – or safer. Apparently, the force of a lightning bolt can spread ten metres from its point of impact.

As my nightmare unfolded, the one sliver of good news was that Nala was sleeping soundly. I'd checked her regularly and she was curled up in her carrier, as happy as Larry.

Camping in the spectacular Turkish countryside.

inner with Jason and Sirem ter the 'day from hell' when ost my passport.

With Jeanie at her shelter near Marmaris, Turkey.

Getting ready to spend the night on a
bench in Sivas, Turkey.

Nala watching the balloons in
Cappadocia.

The tortoise. Cycling through a
village in Georgia.

October 2nd, 2019: celebrating Nala's first birthday, in Tbilisi, Georgia.

Taking a nap in a restaurant in Tbilisi.

With David and Linda in Azerbaijan, October 2019.

'Are we there yet?' Nala's response to the sight of the Azerbaijan border.

So near and yet so far. Looking out over the Caspian Sea in Baku.

Ready to board a train back to Georgia in the race back to Ghost.

Best buddies. Nala and Ghost taking a break together.

With Pedro, Ghost and the other rescue dogs in Tbilisi.

Playing hide and seek near Ankara, Turkey.

Surprise! With my gran for her 90th birthday back in Dunbar, November 2019.

Walking around Plovdiv, Bulgaria, New Year 2020.

Fogbound. Sheltering from the weather in a Serbian field, January 2020.

Exploring the grounds of Buda castle in Budapest, Hungary.

Nala checks herself out on YouTube.

Peek-a-boo on the train through Turkey to Ankara.

Cool cat. Nala knows when to stay in the shade.

Sweet dreams. Dozing in her favourite spot – on my chest.

'I'd spend the day there if I was you,' I told her.

Somehow, I found some extra reserves of energy and managed the final stretch up to the peak, just as the storm passed overhead. By the time I reached the car park at the summit, I was soaked to the skin. My one solace was that the only way forward from here was back down the other side of the mountain. However, it would be another few hours until I reached the next town.

The peak was a popular destination for locals as well as idiotic tourists like me. I recognised several of the cars that were parked in the little car park next to the viewing point. The flatbed truck was there, too. I thought, I've come this far, so I might as well enjoy the view. It was spectacular. By now the storm clouds had moved off to the valley to the side. In the other direction, I could see for what seemed like a hundred miles back towards the coast. I retraced the route I'd taken the past couple of days and was impressed by the progress I'd made.

Someone there spoke a little English and struck up a conversation with me. I explained that I was heading for the city of Konya and after that Cappadocia. He apologised for not being able to offer me a ride down, explaining that most people there were heading back down the way they came. It was disappointing, but by then I'd already given up hope of getting a lift.

I was giving Nala something to eat before heading off again when the lady from the truck approached. Unfortunately, I had no idea what she was saying to me. But then I spotted that the driver, an older guy who I assumed was her husband, had dropped the tailgate down and was busy clearing some space, helped by the son. They were both waving me towards them. I couldn't believe my luck and their kindness. They were offering me a lift.

The back of the truck was filled with a type of gravel, but they had cleared some of it away so that there was room for me, Nala, and the bike at the front, next to the cab and under the tarpaulin roof. Nala looked cheesed off – she'd been enjoying her snooze

– but we soon got ourselves comfortable. The dad cranked through the gears and set us off down the other side of the mountain.

At times I was amazed the guy could even drive. The road was scarily steep, rutted, and full of corkscrew twists. It was dusty, too, even after the brief rainstorm. The truck slid and swerved here and there, but we kept moving. If we ever hit a pothole, or the chassis dragged on a rock, the guy would turn round and give me a toothless grin and a thumbs up: *It's alright, mate, I've got this.* It felt as if he'd driven these roads before, and all I could do was trust him.

By the time we reached the bottom of the mountain almost an hour later, all trace of the storm had passed and we were driving in sunshine again. We must have travelled between twenty and thirty miles when we arrived in a little town. The dad pulled up and dropped the tailgate, and I guessed this was the parting of the ways. I was saying my best goodbye in Turkish when, out of nowhere, he produced a bottle of raki and a couple of glasses. He wanted to toast our little trip.

I couldn't refuse; they'd saved me from a rough ride down the mountain. I raised a glass to him and his family, then swallowed it in one large gulp.

There was enough daylight left for me to make it to another, larger town further down the road. As the family climbed back into the truck and headed off, I hit the road with Nala.

Every bone and muscle in my body was screaming. My calves were agony, and my upper legs were cramping, they were so sore. My triceps were also killing me from pushing the bike up the mountainside. But I could now cycle a little more freely: we were on flat land again.

I reached the next town in about half an hour, as the sun was setting. If I'd been expecting a quiet little backwater, I was disappointed. There was a wedding going on and the streets were alive with people eating and drinking, singing and dancing. The party was in full swing.

I found a spot to park the bike, then got some water from a stream to give Nala. She'd wandered off towards the wedding, where a few little kids in neat white shirts and dresses had already started making a fuss of her.

Some guys were standing outside a small cafe and waved me over. They spoke a tiny bit of English and I managed to explain a little about where I'd been – and where I was headed next. Before I knew it, I had another drink in my hands, this time some Bacardi. I sat with them for a while, watching Nala playing with the kids as we chatted, but I couldn't stay. I'm not normally one to miss a party, but I was absolutely shattered.

I wheeled the bike away from the centre of town, but couldn't find a decent place to pitch the tent. I doubted I had the energy to erect it, in any case. So I parked up next to a comfortable-looking bench in a quiet, wooded area and laid my head there, using a rucksack as a pillow. Nala, as usual, draped herself across my chest, fiddling around for a little as she settled.

For once I was asleep long before her.

16
Team Nala

Three weeks after I'd first spotted it, the sore on Nala's lip was still worrying me.

It was frustrating. One day it seemed to be on the mend and shrinking, the next it would look more livid and raw – and painful – than ever.

I suspected she was picking at it when it was itching and I told her off a couple of times when I saw her putting her paw to her mouth. I felt like a parent scolding their child for chewing their fingernails. She looked at me as if I'd taken leave of my senses.

Against my better judgement, I did some more reading on rodent's mouth.

Some vets said it was transmitted via plastic and so, travelling through the city of Antalya, I had put Nala's plastic bowls in a recycling bin and replaced them with metal ones. I had no idea whether it would help, but it was worth a try.

I also had a couple of conversations online with Sheme and two other vets who followed me on Instagram. The jury was divided. One said the sore was probably nothing to worry about, the other was clear that if it persisted, I should get her back to a vet.

The one thing there was consensus on was that I should give

Nala as much rest as possible. Relaxation was the great healer. That made me feel guilty again; I'd not exactly made life an oasis of calm these past few days crossing the mountains.

So I breathed a small sigh of relief when, towards the end of August, we cycled into the town of Goreme in Cappadocia. I planned on spending a week or so here and Nala could sleep 24/7 if she wanted. I had plenty to keep me occupied.

To start with, I'd finally launched my YouTube channel. If I'm honest, my first video posting was not great. I spliced together some photos and video clips from Santorini, then added a background track. The music was too loud and the edit a bit messy. I knew I could improve. Since then I'd turned footage of our travels in Turkey into two more films of under ten minutes. Each new one was better than the last. My plan now was to put a new video up each Sunday. I'd travel and film during the week, then settle into a hotel over the weekend so that I could edit and upload the video. I would get better with experience, I felt sure.

My biggest asset, of course, was that I was working with a genuine TV star. Not only was she a beautiful cat, Nala was a natural in front of the camera, too. At times I was convinced she was playing to it. This past week, for instance, while we were taking a break from the sun at a roadside cafe, I placed the new GoPro on the floor while she rolled around in the shadows among some dried leaves and stones. At one point she seemed to deliberately roll a stone towards the camera – for dramatic effect. It looked great on film. At other times, she'd press her face up against the lens, looking adorable as she did so. If I hadn't known better, I'd have thought it was her way of saying: *Let me entertain you.*

If that was her intention, then she certainly succeeded. Tens of thousands of people were already subscribing to the channel. Judging from their comments, they seemed to particularly enjoy the 'Nalacam' footage of us riding through the countryside with the star of the show at the helm of the bike. When I added the YouTube audience to the 600,000-plus followers we now had on

Instagram, it was a mind-boggling number of people. I often wondered about them. Who were they? What did they enjoy about our page?

I was fairly certain that, to most of them, we were only a little diversion in their busy days. They loved seeing the latest cute photo of Nala. They'd leave a 'heart eyes' emoji then move on. But there were definitely a large group of people who felt more strongly. They seemed to follow our adventures more closely, sending messages and offering advice and help. I'd begun to think of them as her own personal posse. Team Nala, so to speak.

Naturally, they had opinions. There was plenty of well-meaning advice on subjects ranging from her diet to whether or not I should clip her claws. Others had strong feelings on where I should – or more often, shouldn't – be travelling to next. Given the complications of the world, their opinions differed. A lot. If it was up to the most cautious followers, I'd have wrapped Nala up in cotton wool and put her on a plane back to Scotland long ago.

But what touched me most was the practical help that people were offering. It was genuine and heartfelt, if a little overwhelming at times. The constant flow of gifts had been staunched back in Santorini, thank goodness. But I was still regularly offered bits of equipment for the bike, or clothes for Nala. I said yes to some gear for carrying Nala, and also to a company called Schwalbe in Germany for some super-durable tyres, which I reckoned I'd need when I got into Central Asia and India. But in general, I politely declined or didn't respond. I simply had no way of carrying it all.

The oddest thing, to my mind, was the number of people who offered accommodation if I was 'ever passing their way'. It always amused me, because it seemed so unlikely. What were the chances of me cycling past the front door of someone who'd found me on Instagram? They had to be tiny.

The most inspiring aspect of all was the way people were also willing to dip into their pockets to help the causes I highlighted.

It started with the reaction to Balou in Albania, of course. And then helping Christina on Santorini. It opened my eyes to the potential we had to raise money for deserving causes. As Jeanie said, with Nala at my side, I had a unique opportunity to help people who ordinarily struggled to raise money.

So with that in mind, I spent a lot of my time during my 'week off' looking at ways of capitalising on that opportunity.

To begin with, I finally got around to sorting out the raffle I'd started way back in May on Santorini. I felt bad about not having done so sooner, but I'd simply not had time to organise the draw until now.

It turned out that 13,000 people had bought a £1 ticket. I was amazed.

Having picked the winners and got Galatea back at the pottery shop to dispatch the four bowls, I was now faced with the job of distributing the money. The plan was to give £1,000 each to thirteen different charities. I'd already started work on a list of potential recipients.

Encouraged by this, I also started planning a more ambitious fundraising effort: a calendar, featuring photos of Nala, the profits of which would go entirely to charities. For a while, I was afraid I'd bitten off more than I could chew. I have some skills on a computer, but I wasn't up to the job of designing a calendar. But Team Nala are a talented bunch. I soon found someone who had the skills to make it happen, an American designer called Kat McDonald from New York.

All in all, I felt that I was now doing something worthwhile. I was making a difference.

In the thousands of messages I received, I think there were only one or two that were negative in any way. I knew people were rooting for us and enjoyed following us. I felt I was brightening up people's days and – maybe – shining a small light on the world at the same time. I wasn't getting carried away, though; my little YouTube channel wasn't *National Geographic* or the BBC World

Service. But it was my window on the world, which made it even more important that I start planning the next leg of my journey.

After all, my page was called *1bike1world*.

Having Nala with me meant I had to learn a lot about the art of international cat travel. I now knew that I would need a vet to give her a clean bill of health before moving into each new country. This was a requirement of the international pet passport system. So, even if Nala's lip cleared up completely, I'd have to see a vet in Turkey before heading into Georgia. And I'd have to do the same again before crossing into Azerbaijan.

That would be the pattern from now on. It didn't faze me; the check-ups would be simple. They'd inspect her paperwork, have a quick look at her to make sure she was okay, then sign her off as being fit to enter the next country. The worst thing she might face was a having a thermometer stuck somewhere unpleasant.

Far more daunting was planning the route we were going to take. It was becoming complicated.

A few days back, while stopping for some water and snacks at a petrol station on the main road to Aksaray, I'd bumped into another pair of cyclists, a German couple named David and Linda. They also had a page on Instagram, @*zwei_radler*, the two riders. We shared a coffee and decided to ride along together for a while. It was good to have some company; as entertaining and loving as Nala may be, I missed human conversation at times.

We rode along together for a day, setting up camp in a pretty place called Sultanhani. We explored the town and its ancient, walled mosque, then went out for dinner that night, bringing Nala with us. It was fascinating talking to them. We had a lot in common.

Roughly the same age as me, they had got married a few months previously, back in Bavaria in Germany. Instead of any normal package holiday honeymoon, they'd decided to cycle to Asia.

'We wanted to discover the world not destroy it,' Linda said.

I couldn't have agreed more with that.

After leaving Bavaria, they'd cycled down through Austria, Hungary and through Bulgaria into Turkey.

'We don't have a final destination,' said David. 'We are keeping our options open.'

'As long as we are back at our jobs next March, it's okay,' Linda smiled.

Until recently, David and Linda's plan had been to cycle on a similar route to the one I had mapped out. They were aiming to ride through Georgia and Azerbaijan, before heading across northern Iran into Turkmenistan and then Uzbekistan, to cycle the Pamir Highway, which followed the ancient Silk Road that Marco Polo and others used to travel by land from Europe to China.

They'd planned to cycle through beautiful cities like Bukhara, Samarkand and Khiva, then drop down through the Himalayas into India. Like me, they'd been really looking forward to it and saw it as one of the highlights of their adventure. But as we talked over dinner, they told me they'd abandoned the plan.

'It's late August now and the weather is changing already, I think we will be too late,' explained David. He told me that some of the toughest and most treacherous parts of the highway are at high altitudes and would soon be impassable because of snow. 'It's not the best place in the world to get stranded for an entire winter,' he added.

It came as a bit of a shock. I'd known time was tight, but didn't realise it was that limited. I'd been led to believe it was open until November. David and Linda explained that they were now planning on heading into Azerbaijan and then going south through Iran. They'd aim to get into Pakistan from there, cycling on into India and Burma and Thailand after that.

'That's the plan, anyway,' Linda smiled. 'As you know, plans change when you're on a bike.'

'Tell me about it. Try doing it with a cat,' I said.

We parted company the following morning, but agreed to stay

in touch. Our paths might cross again, we suspected. The encounter gave me plenty of food for thought. But when I began to research it a little, I began to worry.

I'd heard good things about cycling through Iran. It's a spectacular country geographically and it being an Islamic country, I knew they'd be welcoming to Nala. But I also knew there were political problems. When I looked at the Foreign Office's official website, it said that I'd need to enter the country as part of a tour party. There were companies that could sort that out for me, but it seemed like a lot of hassle. It also felt as if I'd be restricted as to where I could cycle. Last, but definitely not least, I'd been told that it was unlikely the hotels on the official tours would accept Nala. She'd have to sleep outside or in a cattery nearby.

That was a dealbreaker for me. So I put a feeler out on Instagram for advice on what to do next. My new army of followers might include some travel consultants. Someone might have an interesting solution.

I received some very useful replies. Someone from Turkish Airlines contacted me about a direct flight to India. I'd always been against the idea of Nala flying, because I couldn't contemplate her going in the hold. I was reassured that Nala could fly in the cabin with me, provided she was in her carrier. But that was going to involve a lot of red tape, which was always a negative for me.

So I decided to press on into Georgia and Azerbaijan. I knew there would be options available, whether by bike, train, plane, or boat. I still had 1,800 kilometres, or more than a thousand miles, to go to the capital of Azerbaijan, Baku, on the edge of the Caspian Sea. I'd work it out. Somehow.

The big attraction in Cappadocia is its amazing landscape. It's like something out of a *Star Wars* movie. The valleys are filled with endless cone-shaped peaks that look like they've been sculpted out of sugar.

The best way to see the view is in a hot air balloon. Every morning at dawn, scores of them rise into the air and float over the landscape. I knew I had to experience it for myself and I made contact with a balloon company. To my amazement, they were aware of Nala and me on Instagram and were very keen for her to accompany me on a trip.

I wasn't comfortable with that idea. I knew Nala didn't like loud and scary noises, and I've never forgotten her reaction on the ferry arriving in Santorini. I'd seen the large burners on the balloons they used here and they were deafeningly loud. I was certain they would freak her out. So I declined their offer and decided to do it on my own. The balloon operators were disappointed and kept trying to persuade me, but I was adamant. I wasn't going to have her terrified for a free balloon ride. I'd pay like everyone else.

A few days after I arrived, I got up at 4.30 a.m. and headed out to the meeting point. I don't have the best head for heights and was a wee bit anxious as I stood in the big basket with about a dozen others. But floating over the strange landscape was an unforgettable experience. The sight of what must have been a hundred multi-coloured balloons filling the pale, pink and blue morning sky was extraordinary. It reminded me why I'd set off to see the world in the first place.

After a week I left Cappadocia, determined to crack on and get up to the Black Sea coast and into Georgia in decent time. I was happy to be the tortoise here and there, but for a wee while now I'd been thinking I needed to act more like a hare. David and Linda had given me a fright.

The good news was that a week's rest seemed to have done Nala the power of good. I'd need to get her seen by a vet before crossing the border, so I decided to pre-empt that by taking her to a clinic here in central Turkey. I asked the vet to take a look at the lip and was amazed at his response. He skilfully managed to open Nala's upper lip and then gave a shrug.

'Nothing there,' he remarked.

'What?' I said, leaning in, not quite believing him.

Sure enough, the lip was completely clear.

The vet went on to give her a thorough check-up, which she passed with flying colours.

'She's a very healthy little cat. You have obviously been looking after her very well,' he said.

The all-clear gave me a huge boost, but I didn't want to risk a recurrence of the lip problem. There was another mountainous stretch of countryside between us and the Black Sea, so I decided I'd take us on a local bus that travelled the route. It would save Nala – and me – the stress of more climbing in the heat.

I was going to catch the bus in the city of Sivas, so I cycled to within a few miles the night before and arrived at the coach station in plenty of time for the ten o'clock departure the following morning. I sat by the bus stop, entertaining myself while I waited and Nala snoozed. Ten o'clock came and went. As did ten-thirty and eleven. There was no sign of the bus.

The coach station offices looked closed, but a small ticket booth had opened mid-morning. I went to enquire. It turned out I'd misread the timetable. My bus was leaving at *10 p.m.*, not 10 a.m. We had another eleven hours to kill.

I've had worse setbacks, so I decided to spend the day exploring the town, taking photos and having a nap in a little park. By early evening we were back at the coach station, settled on a familiar bench outside the depot.

By the time the battered old coach ground to a stop shortly before ten o'clock, I was more than ready for my comfy seat and a sleep. With luck, when I woke up we'd be on the Black Sea coast.

The driver got out and opened the large, enclosed storage area underneath the bus, ready to load people's luggage. There were a couple of other passengers, only one of whom had a suitcase. I'd had plenty of time to pull the bike apart, so I had started to put the trailer and panniers into the hold.

The driver wasn't happy and shouted something in Turkish.

He wanted to do it himself, which was fine. It was his bus. I left him to it.

It walked back to the door of the bus and was about to climb the stairs when I was aware of him shouting at me again.

'Kedi! Kedi!'

Nala was fast asleep in her carrier rucksack. I hadn't even realised he'd spotted her.

I walked back towards him. He was pointing at a mesh-covered box under the bus.

'Kedi.'

I didn't need a translation. He was saying that she was going to have to ride in the storage area. There was no way I was agreeing.

I argued with him as best I could.

'She's asleep,' I said, showing him inside the carrier. 'A-s-l-e-e-p.'

But he was having none of it. Another passenger had arrived and spoke a little English. He spoke to the driver.

'He says the cat will make noise all the way. People won't sleep,' he said, with a shrug of his shoulders.

I waved my arms in surrender. I grabbed my bike and the rest of my gear. I wasn't travelling without Nala at my side.

The driver gave me a look that said: *Your choice, mate.*

It was annoying. I'd taken the trouble to check at the ticket station when I'd bought the seat on the bus a few days earlier and again at the little booth that afternoon. They were adamant that Nala could travel with me provided she was in a carrier.

It was too late to cycle; the roads were unlit and still busy with traffic. There wasn't anywhere to pitch the tent either, so I laid myself out on the bench where I'd been sitting earlier. I'd work out a next step the following morning. I sat there for a while, updating my Instagram and chatting to a couple of people back in the UK. It was getting cold by now, so I got out my sleeping bag. I lay down and nodded off sometime after midnight.

I couldn't have been asleep for more than a few minutes when I felt a dig in the ribs.

Uh, oh, I thought. *My luck's run out at last. I'm being mugged.*

But when I sat up sharply, all I saw was two smiling faces. Female faces.

'Hello,' one of them said in faltering English. 'I see you on Instagram. You no need to sleep here, you come to my house.'

My first reaction was shock. I was in a small town in the middle of Turkey. Less than half an hour earlier, I'd posted the fact that I was going to sleep on a bench here. Now someone had arrived to take me to their home. How crazy was that? And how wrong I had been to underestimate Team Nala – and also the possibility of 'passing by' the home of one of its members.

I explained I'd slept in worse places, but she wasn't taking no for an answer.

We wheeled the bike along the backstreets and soon came to a little house within the town. The lady introduced herself as Arya. Her friend had a really Turkish name that – to my shame – I couldn't pronounce. Arya had even laid out a spread for me. It was a banquet fit for a king; well, it felt like that for someone who was used to grazing on snacks and fruit he found on the roadside. I couldn't thank her enough.

I slept like a king, too.

The next day Arya showed me around Sivas. She was rightly proud of the city and showed me some of its famous sights, including its Islamic colleges, or *medreses*, and its famous Turkish baths. I also managed to get in touch with a guy who offered to drive me to the coast. I'd have to pay this time, but I didn't mind. I wanted to get across the mountains and on to the coast road to Georgia as soon as I could.

By late that afternoon we had loaded all our gear into a smart white van with comfortable seats and air-conditioning. Arya waved us off, blowing kisses to Nala as we headed on to the road. It was odd; we'd known each other for fewer than twenty-four hours, but it felt like we were saying goodbye to a lifelong friend. It also felt like we were saying goodbye to Turkey.

As we wound our way up to the highest point in the mountains, the Black Sea coastline loomed into view, curving its way east towards the border with Georgia. I'd be there in the next few days, I hoped. My time in Turkey would soon be over, but I wouldn't forget it in a hurry.

It proved to me that I'd been right in wanting to see the world first-hand and not trusting a lot of what I read in newspapers and saw on the TV news. It was always too simplistic; too black and white. Too keen to pit people against each other, and to portray people from other religions, races and cultures as being somehow different. I'm not naive. Of course, the world is a complicated place; it is riddled with complex social and political situations. Bad people, too.

But, deep down, I've always wanted to believe that we are all the same. That it's human nature to do good rather than bad. Turkey proved me right – and in spades. So many people offered us the hand of friendship. I now thought of them as extended members of Team Nala. Jason, Arya, the family in the truck, and many others – they'd done it because it was their natural instinct to come to the aid of someone in need.

I doubt I will learn a more heartening lesson, even if I circle the world twice over.

17
A Different World

I'd just swerved the bike to avoid a stray goat in the road when I noticed Nala sitting bolt upright, her paws gripping the handlebars. Her ears were pricked and she was flicking her head rapidly from side to side, as she did when something had caught her eye. I soon realised it wasn't the goat.

A short distance ahead of us, an elderly lady in a black smock and bonnet was shouting and waving a stick at two large white cows, trying to steer them towards a small cottage on the edge of a wood. The cattle weren't the most obedient creatures. They had stubbornly come to a stop yards from the front door and were mooing loudly, as if in protest. The lady was having none of it and started shouting louder than ever, smacking one of the cow's backsides with her stick for good measure. It did the trick. Moments later, the two cows were ducking their heads under the beams and squeezing through the tiny doorway into the cottage. The lady wasn't far behind them, closing the door after her.

I couldn't help laughing out loud.

'Maybe she's invited them for tea,' I said, ruffling Nala's neck as I pressed on down the road.

I'd been in Georgia for a few days now. Sights like this added to my feeling that I'd entered not only another country but a different world, maybe even a different era.

The first clue came at the border, where the tailback of trucks waiting to cross went on for miles. I cycled past most of them, but was still forced to wait an age. When I finally got to the front of the queue, I discovered a heavy military presence and a phalanx of border guards, who were going through everyone's paperwork with a fine-tooth comb. At one point, Nala climbed off the bike on to my arm and tried to catch the eye of one of the officials, a young guy. For once her charm didn't work: he walked away. His colleagues spent what must have been fifteen minutes passing our passports around and talking animatedly, before waving us through with nothing more than a grunt from one of them.

As I eased my way down the road, my first glimpses of the Georgian landscape reminded me of Albania. It wasn't that long ago that the country had been part of the old Soviet Union and every now and again I passed an austere, official-looking concrete building that had been left to crumble and decay. The houses and villages that we passed through looked equally battered and weather-beaten. Most were in a very shabby state. So too were the roads, which were filled with potholes and craters that made riding difficult.

It was made harder by the drivers, who had no sense of keeping to a lane. Nala and I were lucky not to get knocked over, as vans, cars and trucks were passing so close that even Nala shot the drivers a withering glance. I soon decided to take the smaller and safer country roads wherever possible. The Georgian countryside was beautiful; a lush landscape of rolling hills and picturesque river valleys, with imposing mountains in the distance. But the deeper I travelled into it, the more it felt like I was stepping back in time – or into some strange land, straight out a folk tale.

The old lady and her cows were typical. Animals were everywhere and seemed to be living cheek by jowl with their owners. Goats, cattle and scrawny, underfed-looking horses were often wandering in the middle of the road. There were dogs, cats, geese, pigs, chickens, you name it. It was like a zoo – as if the roads were part of people's farms and smallholdings. Except no one seemed to care much for any of the animals that were among them. Most were in a terrible state.

Of course, some will argue that these people were too poor to worry about animal welfare, and I can never know what it's like to be in their shoes, but you don't need money to be kind to another creature. I found it upsetting, and it took the shine off the beauty of the place.

I was heading for the first sizeable town, the Black Sea resort of Batumi. We got there late in the afternoon, in time for a torrential rainstorm. Georgia may be a poor country, but as I'd already discovered, the most hospitable people are often those who have the least. I checked into a small hostel where the guests ate communally in the evening. They took me in as if I was their long-lost brother. Within moments, I was being offered my first glass of vodka.

That was the beginning of the end. One glass led to another, then another and another. I got so drunk that I could barely move the next morning. My plan was to take a week or so to cycle through to the capital, Tbilisi. I was going to pick up the new tyres that Schwalbe had dispatched to me, but the hangover set me back almost instantly. It took me a day to get over it.

When I did manage to get back on the bike, the ride through the Georgian countryside continued to be an eye-opener.

In one small village, a tiny car was revving its engine like crazy as it tried to tow a lorry five or six times its size. I've never seen anything like it and I've no idea how the driver thought it was even possible. Nearby a group of children were playing with a battered old basketball, laughing and cheering each other on as they lobbed it at an old bucket they'd hung on a tree. They reminded me of the kids at the refugee camp all those months ago. Like them, they didn't need the latest, most expensive gear to have fun.

Animals were all around us again. Every home, it seemed, had a small menagerie of goats, chickens and donkeys. Dogs of all breeds, shapes and sizes were everywhere. Most must have been strays; they seemed to wander the lanes and fields freely, as if in search of their next meal or night's sleep. It was heartbreaking to see. A couple of hours into the journey, one of them – a leggy, flappy-eared character with a spotty brown-and-white coat – latched on to us.

It reminded me of Teal, the pointer we'd had back home in Dunbar, except in this case the dog was in an awful state. It was little more than skin and bones. It was walking along about twenty or thirty yards behind us, its head hung low, its gaze fixed upon us. I stopped to say hello at one point and gave it a small snack, and the dog gobbled it down in one gulp. The poor thing probably hadn't seen food in days. I imagined it would head back towards whatever it considered to be home, but instead it stuck with us for miles and miles. It was only when I arrived back at the main highway to Tbilisi that I managed to shake it off.

As I slipped back into the traffic, I couldn't help turning round, only to see the dog still standing forlornly at the edge of the road. That was it. I couldn't stop thinking about it for the next twenty miles of my journey.

The idea of going back to rescue the dog didn't leave me for days.

Seeing these poor creatures made me cling even tighter to Nala. She'd been in the same boat as them; left to her own devices, alone and helpless. Thank God I'd found her and was able to give her a healthier, safer life.

The worries I had about her during the early days and again in Turkey were a thing of the past. There was no hint of the sore on her lip and the vet in Turkey had declared her as fit as a fiddle. If only I could have said the same for me and my bike.

I don't know if I picked up something on the road, or inside the dusty rooms in the cheap hostels I'd stayed in, but I developed a weird itching in my eyes. I wasn't able to resist rubbing them, which made it worse. I was about midway through my journey across to Tbilisi when it got so bad, I could barely open them at all. Cycling with your eyes closed isn't recommended – especially on Georgian roads – so I had to abandon one day's cycling after only twenty miles or so.

I also kept getting punctures. It wasn't surprising given the state of the roads, but it was an indication that the old tyres were on their last legs. On top of all that, I noticed a crack on my disc brake; it was badly bent. Luckily there was a Trek garage in Tbilisi, as I suspected my bike would need a serious service.

So, when the weather took a severe turn for the worse a few days out from Tbilisi, I made a decision. I'd spotted that we were near a railway line with a regular service into the capital. As we made our way to the station, the skies turned almost black and I heard the loudest thunderclap I've ever heard in my life. We hopped on the train mid-afternoon, in the nick of time.

The rain was falling so hard that I barely got a glimpse of the Georgian countryside. But Nala was entertained. She kept pawing at the raindrops streaming down the window of the train carriage as we trundled along.

We arrived in Tbilisi by early evening and headed for a small apartment I'd rented high on the hill overlooking the city. Nala loved it; there was space for her to run and jump around. We could play there for hours on end without anyone bothering us – or vice versa. It was also safe and secure enough for me to leave her there if I needed to pop out on my own.

During my first couple of days there, I took my bike for a service and collected the new tyres. They had an extra protective layer around them so, in theory, when I cycled over glass or nails, or anything else sharp, that extra skin would protect the inner tube. It almost felt like having a new bike.

The flat was a good base for me to catch up on my online

activities. By now I had started distributing the money I'd made from the raffle of pottery. It wasn't easy. There were so many deserving charities that it made my head spin. So, when it came to picking the first recipient, I was led by my heart as much as my head.

I was about six years old when my grandad and I planted a tiny holly tree in his garden, not far from our house in Dunbar. I've never forgotten him telling me how important trees were to the planet and how they provide oxygen, store carbon, stabilise the soil and even provide homes for some of the world's wildlife. When he passed away, my dad and I dug up the holly tree and replanted it in our own garden, where it remains to this day.

So in memory of my grandad, I gave the first £1,000 to a charity called One Tree Planted, a non-profit organisation dedicated to global reforestation, which planted a new tree for every one dollar donated. It felt good to think that there were going to be more than a thousand new trees in the world, all dedicated to the memory of my grandad.

The first choice was the hardest. After that, some of the other charities were no-brainers, particularly those whose leaders had helped me along the way, such as Jeanie, Lucia and Christina. I knew they could do a great deal with £1,000. I also gave money to other environmental causes I felt strongly about, such as a coral preservation fund in Australia.

I was under no illusions that I would have raised a penny without Nala. So, in the first week of October, I found her the best tin of tuna that Georgian money could buy. She deserved it, not only for giving me such a great opportunity, but also because it was her birthday. She'd reached the ripe old age of one, according to the date that the vet in Montenegro put in her passport.

It was a glorious end-of-autumn day and the temperature in Tbilisi was close to 70 degrees. We celebrated with a long walk around the city, drawing admiring looks from the locals as Nala sat on my shoulder or toddled along at the end of her lead.

We spent an hour in one of Tbilisi's many pretty parks. Nala had a ball playing in the flower beds and clambering up trees. Watching her having fun, I couldn't help dwelling on how much she'd grown since we first met. She was four or five times the size of the little kitten I'd found at the roadside, but it was her personality that had matured the most. I read somewhere that the first year of a cat's life is the equivalent of fifteen years of a human one. That made sense to me. She may not have been as wild as I was at that age, but she certainly had many of the traits of a teenager.

At one point in the park she tried to charge after a flock of pigeons that was being fed by an elderly couple. She was on the end of my extendable lead, so I was able to stop her and lead her away. But she kept trying to wriggle out of her collar until we left the park. Her meows of protest were so loud, they could have been heard halfway to Moscow.

The walk in the park tired her out, and we went out to lunch at a nice restaurant in the old town. Nala decided to fall asleep in a flowerpot in the middle of the tables, much to the amusement of the locals and tourists who were eating there. I laughed to myself. Even when she wasn't awake, she had the ability to entertain people, to be the centre of attention. By the time I'd walked her back up the hill to our temporary home late in the afternoon, she was exhausted. I ended the day's celebrations on my own, sipping a beer and catching up on our Instagram, where there were hundreds of birthday greetings.

I felt quite emotional reading them. The outpouring of messages brought home to me how much of an impact Nala had made, not only on me. It also made me think about what might have been if our paths hadn't crossed. Where would I be now? Lounging on a beach in Thailand, or Australia? Or would I be back in Dunbar, my journey over, my hopes of changing my life dashed? Fortunately, these were hypothetical questions. I was here, with Nala, relishing every minute of our unexpected adventure together and feeling rather proud of how much we'd achieved – and what more we could do together.

I was also thinking more about my next steps. Travelling across Iran was still tempting, but I couldn't find a way around the difficulties. The political situation had become even trickier and if I upset the authorities for any reason, and maybe ended up in an Iranian jail cell, it would be all my fault. This was not a fate I could contemplate for Nala either.

I was disappointed at the thought of not following the Pamir Highway and considered waiting out this winter and heading there in the spring, when the mountain passes became accessible again. But where would I stay holed up for five months?

I liked Tbilisi, but there wasn't enough for me to do here. If I headed back to Istanbul and Turkey, I'd give myself more options. I could stay there a while then cycle up the Black Sea coast into Bulgaria and Romania, and perhaps even further north in eastern Europe. There was also the option of flying to India – or somewhere else. I found it all too much to process. So I decided to live in the moment and stay focused on my main priorities: looking after Nala and getting on with my new job.

My plans for a Nala calendar had suffered a few ups and downs. The company that originally said they'd handle the distribution had pulled out. They simply couldn't deliver on the global scale I had planned. So I turned back to Team Nala on Instagram and was busy looking for an alternative, one of which relied on friends and family back in Scotland. I was sure I'd find an answer. The list of charities I wanted to help seemed to get longer every day.

My YouTube channel was also growing nicely. It had even begun earning me a little money, which was welcome. It was interesting to work out which videos people liked best. They seemed particularly to love the little shorts I'd made of me and Nala play-fighting while I cycled, along with another one of us snuggled up in the hammock together, back in Santorini when I'd been recovering from my leg injury; this video was getting tens of thousands of views. But people also enjoyed the slow-moving cycle tours through the countryside. It was a great way of seeing parts of the world that they would never get to visit otherwise, like virtual travellers. The ride through Georgia turned out to have an ideal landscape. As I prepared to leave Tbilisi, I was hoping Azerbaijan would match it.

I set off early one morning towards the end of October. The cycle into Azerbaijan was around thirty to forty miles, but this part of the journey was on good quality roads. For a large section of the trip, I was cycling on a high-speed motorway with something resembling a bike lane on the inside. With my new tyres in place, I was soon making good progress and was within half a kilometre of the border by mid-afternoon.

I was congratulating myself on the achievement and had pulled over to get our documents ready when I spotted something at the side of the road. At first it was hard to make it out; it was little more than a flash of white. But by now my instincts were so well tuned, I knew almost immediately what it was. A dog. Even from a distance, I could tell that it was in distress by the way it was twitching. I jumped off the bike and took a closer look.

I'd seen some bad things travelling through Georgia, but this was a woeful sight. It was a small, off-white puppy, no more than a few weeks old. The poor thing was skinny, dehydrated and struggling to keep its eyes open. It barely had enough energy to wag its tail, and I felt it had already given up its fight for life.

My reaction was automatic. I couldn't leave it here to die; I had to do something. But I was soon feeling a sense of *déjà vu*. It was the same feeling I'd had at the Montenegrin border ten months earlier. My mind was flooded by competing thoughts. What was I going to do with it? Where should I take it – and how?

Repeating the trick I'd pulled off back in Bosnia wasn't on the cards. Apart from anything else, I was already within view of the border crossing. Someone may have been monitoring me even at that moment. I also didn't have any paperwork. If the Georgian border had been anything to go by, border authorities in this corner of the world were very strict. They'd almost certainly confiscate the dog.

It didn't take me long to work out what to do.

I leaned down and picked up the puppy as carefully as I possibly could. It yelped with pain and wriggled a bit as if trying to escape. I reassured it and placed it inside the backpack carrier.

Nala was shooting me vexed looks.

Who have you picked up this time?

I gave her a stroke on the back of the head.

'Sorry, Nala, bit of a U-turn,' I said, as we spun round and started on the trip back to Tbilisi.

It was early afternoon by now and I wanted to get back to the city in time to catch a vet. The puppy didn't look as if it would make it through the night without treatment of some kind, but the ride seemed to take forever. Hills suddenly appeared where there hadn't been any on the journey out that morning.

By the time I hit the outskirts of Tbilisi late in the afternoon, I was sweating buckets.

I'd found a twenty-four-hour vet and rung ahead to say I was coming. The staff took one look at the puppy and sprang into action.

Within minutes it was hooked up to an IV drip to get fluids flowing back into its body. A vet and nurse then took the puppy into an X-ray room. I watched with the two women as they looked at the black-and-white images on a monitor.

'His bones aren't in good condition, looks like he may have weak joints,' the nurse was able to tell me in English.

The vet then said something in Georgian.

'She can't be one hundred per cent certain, but she also thinks he has probably eaten something he shouldn't have,' the nurse told me.

She explained that the vet wanted to keep the dog for further tests over the coming days.

'And then what?' I asked.

'Depends on whether he's got a home to go to,' the nurse answered, rather ominously.

That made my mind up for me.

'I'll give him a home, don't worry about that,' I said. 'But I'm going to cycle to Baku in Azerbaijan. Can you hang on to him until I get back?'

'How long will that take?' the nurse said, looking a little doubtful.

'Ten days probably?'

'Okay, ten days. But after that we will have to find somewhere for him to go. There aren't many shelters for animals in Tbilisi.'

'Don't give him away,' I said. 'I'll be back, I promise.'

We swapped phone numbers and agreed to stay in regular touch.

'I'll see you soon,' I said to the little pup as I headed off.

I hit the road again the next morning with a new purpose.

18
Tea Time

According to my map, it was 300 miles or so from the Georgian border to Baku on the edge of the Caspian Sea. I'd need to do that journey in a minimum of a week, an absolute maximum of nine days. That would allow me to get the train back to Tbilisi in time for the puppy's discharge. I daren't leave it any later; it wasn't realistic to ask a vet to keep looking after him. It was going to be tight. I'd have to hope the roads and weather were kind to me and that Nala and I remained fit and well. To begin with, all went to plan.

After the experiences of Albania and Georgia, I was worried about crossing into another post-Soviet country, but all the Azerbaijani border guards wanted to do was chat, stroke Nala and take photos of her. They waved us off as if we were departing relatives. It was a good omen. The skies were blue and the temperature unbelievably warm for October, but the landscape wasn't the most thrilling on earth. I could see impressive mountains in the distance, but the main road took me through a flat and arid countryside.

For a large part of the trip it also felt as if I was cycling through a giant construction site. Heavy plant and machinery were everywhere. I knew Azerbaijan had huge gas and oil wealth, and it seemed the whole country was being rebuilt on the back of its newfound fortune. The most common sight on my journey was something more mundane, however. Azerbaijan might have been rich in gas and oil, but its people seemed to be fuelled by something else – tea.

I'd started drinking the dark, deep red tea they serve in this part of the world in Turkey and again in Georgia. It usually came in small, pear-shaped glass cups and was slightly bitter to the taste. I usually took it with some sugar. I quickly discovered that drinking this tea was an obsession here in Azerbaijan.

Wherever I stopped, people insisted I sit down and share some. They'd hurry off to the kitchen then reappear with a tray of tea – or *chay* as they called it – accompanied by sweet cakes or breads and jams. It was an important custom, a way of showing that their culture extended the hand of friendship to strangers. The only problem was time. On the first two days of the journey I took tea four times, but after that I had to decline politely. If I accepted every invitation, it was going to take me ten weeks instead of ten days to get to Baku.

By the third night – a Sunday – I'd done well enough to spend a relaxing day in a hotel in the town of Ganja, getting my latest YouTube video edited and published. I'd started making longer films now, up to twenty or even thirty minutes. I had a lot of footage that I thought people would enjoy. Why not? It seemed to be going down well with the followers, too.

I was also continuing to give out my weekly £1,000 to charities. I'd stayed in touch with the vet in Tbilisi, so I was thinking of animal charities again. That week I chose a charity called Street Cats of Oman, run by a lady called Lesley Lewins, who was originally from England, but had moved to Oman when her husband got a job there. She was doing something similar to Lucia and Jeanie in Greece and Turkey. I gave another thousand to Animal Aid in India, a charity that taught children and communities to think more about animal welfare. I wished I could find something similar in Georgia. If ever a country needed better educational resources for the care of animals, it was there.

I felt the donations were beginning to make an impact now. I also managed to do some work organising the calendar, choosing photographs, as well as images by a brilliant Canadian artist, Kelly Ulrich, who had started drawing daily cartoons based on our adventures. If I could sell ten or twenty thousand copies, I was excited at what we could achieve in terms of making a difference.

I was back on the road on Monday morning and was soon halfway to Baku and halfway through my ten days. I'd pulled into a petrol station to top up with water when I recognised the two bikes parked outside.

Sure enough, when I went inside the small store, I found David and Linda. They had taken a very different route to me since we'd parted back in Turkey. After crossing into Georgia, they'd headed north of Batumi into the Caucasus mountains, close to Europe's tallest mountain, Mount Elbrus, across the border in Russia. From there they had cycled their way down the spine of the country and were now cutting across Azerbaijan, towards Iran. I was pleased to discover we were about to be sharing the same stretch of road for a hundred miles or so before separating about fifty miles from Baku, where they'd head south rather than staying east.

Nala was pleased to see them again, too, especially Linda with whom she'd formed a lovely bond. As we made camp on the first night, the pair were soon playing together.

I was doing well timewise and was hopeful I'd get to Baku with room to spare before heading back to Tbilisi. So the prospect of cycling with Linda and David was appealing. It would make my days much more sociable, especially as we were being offered tea and cakes at almost every turn.

Shortly before the major junction where we were going to split up, we stopped at a roadside cafe. We'd decided to have a final meal together before heading our separate ways.

It wasn't the smartest-looking place. I ordered an omelette and was suspicious of it as soon as it arrived. It was grey and watery, and it tasted of nothing much at all. But it was food, so I ate it with some bread and washed it down with some tea – naturally – before pressing on.

By the time we were ready to say our goodbyes, I was already feeling queasy. I was sure I'd eaten or drunk something that had disagreed with me. The omelette was the prime candidate; I should have listened to my instincts and avoided it.

I must have looked rather green around the gills, because David and Linda kept asking me if I was okay and whether I wanted them to stay with me.

I reverted to my old, macho self.

'I'll be fine, it's nothing,' I said, waving them off, but knowing deep down that was far from the truth.

It wasn't long before I was feeling even worse. After no more than a couple of miles, I hit a brick wall. My legs felt like they were weighted down with lead; I felt dizzy and the sweat was pouring out of me. I'd cycle a short distance then have to stop again. The countryside had suddenly become rugged; there was little sign of life in the grassy hills to the side of the road. So I kept stopping, guzzling down water, and trying to vomit. Nala had already picked up on my distress and was perched in her carrier, staring at me each time I bent over a ditch or a stream trying to force myself to throw up. I was so weak, I could barely pedal the bike forward. A couple of times I wobbled and nearly fell off.

Is this it? I thought to myself. *Is someone going to find me dead in a ditch in the middle of Azerbaijan?*

The road had taken me through empty stretches of country, but I was now travelling through the most barren part of all. There wasn't a soul in sight. I looked at the map on my phone and tried to find the nearest hotel, but it was forty miles away. It was soul-destroying; I had no idea how I was going to carry on. A few miles further down the road, I collapsed on the floor outside a derelict building.

Thank God I had Nala. Even though she had a big, extendable lead that allowed her to stray twenty or thirty yards away, she had no interest in exploring. She lay next to my neck, purring away and giving my forehead the occasional lick.

I slept for an hour or so, then somehow gathered my strength to continue the journey, but it was the hardest forty miles of my life. When I finally checked into the hotel, I jumped into the shower, fully clothed. Crocs and all. I was a mess: it really wasn't pleasant.

I came out of the shower and collapsed on the bed, the room spinning wildly around me, my mind and body overheating with fever. That night was the longest and most difficult of my trip so far. I was delirious and had the weirdest dreams. I had visions of the little white puppy being thrown back onto the side of the road; of Nala running behind my bike as if in distress. My mum and dad back home. Weaving my way along an endless road filled with huge trucks that kept threatening to hit me; falling off the bridge in Mostar and never landing again.

It was scary stuff at times, but I had Nala with me to comfort and calm me down. She lay there throughout, nuzzling up against me, purring quietly as if offering reassurance. Whenever I woke up from another crazy dream, I saw her face. She was my anchor to reality. On Santorini I convinced myself that I'd been imagining it, but there was no doubt this time: she knew I was unwell. So she had to be Nurse Nala once more, my furry little angel, watching over me. I've never been more grateful to have her.

The night seemed to stretch on forever. I was up and down out of my bed, running to the toilet, sticking my head under the shower. It was deeply unpleasant, but it did the trick. By the following morning, I'd begun to improve. I was able to sip a little water and eat a couple of bits of bread. I'd definitely turned the corner.

Nala seemed to spot this, too. She left my side now and started bouncing around the room, as if asking me to play hide and seek with her.

'Not now,' I said, sorting her some breakfast instead. 'We need to get back on the road.'

A little wobbly, we were back on the highway to Baku early the next afternoon. My legs still felt heavy and my breathing wasn't great, but I was on the move again. After the darkest night of my journey so far, I was grateful for that.

The sickness seemed to have drained my energy. As I cycled into Baku the following day, I felt strangely flat. It was partly the illness, but also frustration at knowing I'd have only twenty-four hours here before heading back to Tbilisi. The whole object of this journey was to explore and experience interesting new places. I would hardly scratch the surface here, which was a shame. Baku is an impressive place, a mix of flashy, modern architecture and historic old buildings overlooking the Caspian Sea.

I booked us a room in a modern tower with a breathtaking view. As evening fell, the city's spectacular skyscrapers were transformed into a multi-coloured light show. The city looked like something out of *Blade Runner*, or some other futuristic movie. Nala and I stood on the balcony of our room taking it all in.

I felt a curious cocktail of emotions.

I was thousands of miles from Dunbar. I'd reached the gateway to Central Asia and the subcontinent of India and the Far East beyond. Except I hadn't, of course; it wasn't a gateway to anything. I was not going any further along this route.

It was weird. A new and exciting part of the world was tantalisingly close. If I'd smuggled us on board one of the ships I could see leaving the harbour this evening, I'd be on the other side of the Caspian, in Turkmenistan, by the following night. But I'd probably be in a jail cell soon after that. I didn't have a visa and it didn't bear thinking about what might happen to Nala.

It was a similar story when I looked to the south-east and Iran. Some of the giant gas and oil tankers that were lined up on the docks were probably bound for Iran's northern coast. But even if

I had a visa, I'd still face all sorts of risks. I'd read about two cyclists, one from the UK, the other from Australia, who had been chronicling their ride from London to Sydney on YouTube. They had been locked up in one of Tehran's most notorious jails after sending up a drone. It had been totally innocent, but they'd not realised they were near a military base. Without a fair trial, it was anyone's guess how long they'd be stuck in their cells.

Now wasn't the time to overthink where I'd head next. The ten days were almost up and I must head back to the puppy in Tbilisi. I had to make the most of my short time in Baku. I took a whis-tle-stop tour of the city late that night, then dashed madly around the following morning, first to find a vet who would certify Nala as being fit to go back into Georgia. Nala was an old hand at this now and let the charming young vet go through her checks without the tiniest meow of protest.

'Good girl, Nala,' the vet kept saying.

She'd stamped and signed Nala's passport within half an hour. It gave me enough time to do a little more sightseeing, taking in the beautiful old city and seafront.

By early evening, we'd arrived at Baku's gleaming new railway station. Our train, a grand-looking Pullman straight out of *Murder on the Orient Express*, was waiting for us. So too was a stern-faced lady in a plum-coloured uniform.

She had the look of someone who wouldn't bend the rules.

I'd been worried about getting all my gear on the train. There was a lot of it. But as she inspected our tickets, the lady seemed more concerned about Nala. She kept pointing at her and waving her arms around animatedly.

Not again, I thought. *If I miss this one, I'll never get back in time for the puppy.*

Fortunately, another couple of inspectors appeared. One of them, a young lad of no more than eighteen, spoke fairly good English.

'It's okay,' he said, after a heated conversation with the lady.

'She didn't see a cat travelling on a train here before. But you have a ticket, so please go ahead and board.'

'Thanks,' I said, giving him a pat on the back.

He leaned in to me, a conspiratorial look on his face.

'But be careful,' he whispered. 'She is travelling on the train. So you'd better stick to the cabin in case she catches you wandering the corridors. The cat is not allowed to do that.'

A few moments later, we were settling into a cosy little two-bed cabin that would be our home for the twelve-hour journey ahead. The sun was setting as we pulled out of the station and eased our way out of the city. The rhythmic rattle of the train soon sent Nala into a deep sleep.

In the fading light I could more or less recognise the main road into Baku. It felt odd; it was only yesterday that I'd cycled down it. At another time, in different circumstances, I might have felt disappointed or frustrated to be doubling back on myself. But I didn't feel that way at all. For a start, if the past few months had proven anything, it was that my journey around the world wasn't going to look like anyone else's. It wasn't going to run in a straight line or conform to some well-trodden path. I was travelling in Nala's world, after all. And as long as I had her beside me, that was fine. We'd look after each other.

I also knew I had a powerful reason to head back to Tbilisi. I was a man on a mission. The puppy's fate depended on me, and I couldn't let it down. As the train picked up speed and I looked out into the inky blackness of the Azerbaijani countryside, I had no doubts. I wasn't going backwards; I was moving in the right direction.

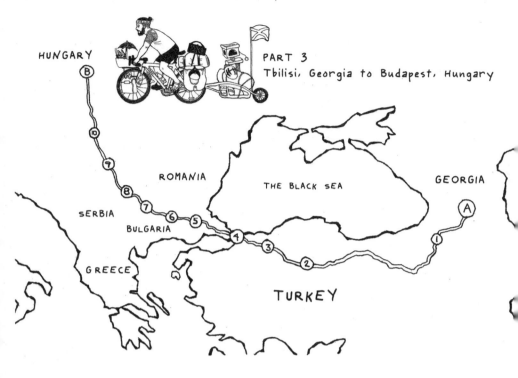

PART 3
Tbilisi, Georgia to Budapest, Hungary

MAP 3

A. Tbilisi, Georgia

1. Kars, Turkey

2. Ankara, Turkey

3. Sakarya, Turkey

4. Istanbul, Turkey

5. Svilengrad, Bulgaria

6. Plovdiv, Bulgaria

7. Sofia, Bulgaria

8. Niš, Serbia

9. Velika Plana, Serbia

10. Belgrade, Serbia

B. Budapest, Hungary

PART THREE
THE ROAD AHEAD

Georgia–Turkey–Bulgaria–
Serbia–Hungary

19
Ghost

I had received a reminder from the vet to collect the puppy soon after I boarded the train the previous evening, so when we pulled into Tbilisi station early the next morning, I wasted no time and headed straight for the apartment I'd rented in the old town.

I dumped my stuff, left Nala safe inside with plenty of food, then made a mad dash to the surgery.

The staff greeted me with knowing smiles.

'We were expecting you,' the English-speaking nurse said, holding up her phone. She showed me the Instagram photo I'd posted of Nala looking out the window of our cabin, as we'd pulled into Tbilisi in the dawn light a couple of hours earlier.

She then waved me through the reception and into the back of the surgery.

'We've got him in the kennels upstairs, wait here and I'll bring him down.'

She returned a few minutes later, the puppy nuzzled into her arms, looking so much healthier than when I'd left him. His coat was less matted and cleaner and even his eyes seemed clearer and full of life.

'You'd hardly recognise him,' I said.

'He's responded well to the medication, but he's still dragging his front paw and the joints in his back legs are weak as well,' the nurse said, nodding.

She put him down on the ground and for a moment we watched the puppy playing. He was wagging his tail and doing a little dance, as if asking for some attention.

'So, what are your plans?' the nurse asked.

I'd already had a discussion with my sister Holly back in Dunbar. She and her partner Stuart had been looking for company for their dog Max. If we could get the puppy back to the UK, they'd love to adopt him.

'Going to try to get him to a home in Scotland,' I said. 'So we will need to get him a passport.'

'Okay. Well, when he is a bit stronger, maybe ten days or so, we can give him the first injections,' she said. 'We can also get him microchipped.'

I felt a sense of *déjà vu* once more.

'Aye. And then a rabies shot when he's three months old.'

She nodded and smiled.

'You've done this before, obviously. With Nala. But it's four months for a dog,' she said. 'So we would give it to him early in the New Year.'

She grabbed some paperwork and I started filling in the forms that I needed to be able to take him away. This time, I'd already thought of a name. When I found the puppy, I'd been listening to the rapper Yelawolf. Somehow his name had sparked a memory of *Game of Thrones* and Jon Snow's white direwolf, Ghost.

'Hello, Ghost, let's take you home,' I said, as I scooped him up and put him in Nala's rucksack carrier.

I headed back to the flat via a pet shop, where I picked up some squeaky toys, a bed, some food and a pair of eating and drinking bowls.

Nala normally came charging up to me when I got home, pressing against my leg and demanding I pick her up for a cuddle.

But back at the flat today she was more interested in the other arrival, sniffing at Ghost as if trying to confirm that he was the same dog she'd met back on the road a week and a half ago.

Ghost, by contrast, wanted to have some fun. As soon as I put him down, he made a few high-pitched whines, inviting Nala to play with him. She was wary at first, but as the afternoon wore on, the gap between them became smaller and smaller. By early evening they were lying on the floor, nipping and rubbing up against each other like two kids rolling around in a sandpit in the park. It had been a hectic week, but seeing them both playing happily together made everything worthwhile.

My aim was to spend the next week or so trying to settle Ghost into life outside the vets, then to find him a home for the three or four months he had to wait before heading to Scotland. I needed to be back on the road to Istanbul to get the plane to India. Of all my options, this was now looking like the best, and possibly only, one.

In the meantime, I hoped to give Ghost as much attention and care as possible. That was the least he deserved.

I could tell that his front paw was still giving him pain. He'd wince and squeal when he tried to put weight on it. The movement in his back legs wasn't great either. His legs kept sliding out from underneath him as he shuffled along on the shiny, slippery floor of the apartment. It was as if he was on an ice rink. I believed his strength would improve with more exercise, so I threw toys around the flat for him to chase. Nala joined in, too, tearing after everything like a dervish.

I could tell Ghost was a good-natured dog, but like Nala during her first days with me in Montenegro, it was clear that he'd been through something traumatic. Every now and again he'd get anxious, freezing and looking around as if in fear of something. He was petrified of the squeaky toys I'd bought him. He'd bite into them, then jump backwards with a start when they made a honking noise.

I've been around dogs enough to know that feeding time would be a big test. They can get aggressive and territorial about food. So when the time came, I made a point of separating the two of them, placing Nala's bowl on the windowsill downstairs and feeding Ghost on the floor at the other end of the large room. I didn't want to risk a flare-up.

Ghost looked at the food I'd laid out in the bowl almost in disbelief. But the moment he realised it was for him, he snarled and growled at me. The message was clear: *This is mine, stay away.* He emptied his bowl in two seconds flat.

I was delighted, but also relieved. If Ghost had gone for Nala during her feeding time, their blooming friendship would have been over before it began. Instead the bond got stronger, and sweeter, to see.

As I sat on the sofa watching a movie that evening, the two of them lay close together at the other end, Nala fixated on Ghost's tail as he flicked it idly from side to side.

Despite their growing closeness, I decided to separate them again at bedtime and created a little sleeping space in an alcove for Ghost, blocking him in with some chairs and a clothes horse. I didn't want him roaming the flat during the night, ripping things to pieces. Meanwhile I'd taken Nala up into the mezzanine area with me.

She clearly didn't approve. I woke up in the middle of the night to discover she wasn't on my chest where she normally slept. I looked downstairs and saw that she'd positioned herself on one of the chairs by Ghost and was snoring away contentedly. I almost felt jealous.

With the following stage of my journey now all planned, the next part in the jigsaw was finding a place for Ghost to stay while the paperwork was sorted.

He could then travel to Scotland in the New Year.

As I'd been warned, there was a real shortage of shelters in Georgia, but I'd been approached by a guy via Instagram who sounded almost too good to be true. He was Spanish and his name was Pablo or @bikecanine on Instagram. Along with two other guys, he was also cycling the world west to east, in his case with a dog named Hippie. His friends had rescued two small dogs in the Georgian countryside and together they were running a mini-sanctuary during the winter, while they waited for the route into Central Asia to open up again.

They were staying in an apartment with a nice open yard. I felt like Ghost would fit in perfectly. I headed over there after a couple of days recovering from the dash back from Baku.

Pablo and his pals were good guys, and we hit it off straight away. The two small dogs looked uncannily like Ghost. They were both whiteish in colour and around the same size. They could easily have been mistaken for siblings, which made me feel even more comfortable about leaving Ghost there. Pablo and I agreed that he would look after him until the New Year. There was still money in the fundraiser page I'd started for Balou to pay for Ghost's paperwork to be sorted, for when Holly and Stuart adopted him. But Pablo refused to take anything. He had his own charity page, he told me.

The more I got to know the guys, the more I was convinced the arrangement suited us all perfectly. The two dogs rescued by Pablo's friends hadn't been in as dreadful a state as Ghost when they were found, but one did have a big infected area on his side. They were due for their first injections as well, so we decided to take all three of them to the vets for their vaccinations the following day. That way, all three would be on the same timetable.

We met at the vets the next day, but it wasn't as straightforward as we'd hoped. The pup with the infection needed a lot of treatment to clear the problem. Ghost in the meantime needed to get his bloods taken; the vet didn't like the look of his skin and wanted to test it further. But by the time we got home, we knew what needed to be done in the coming weeks. I felt that I'd set Ghost on the road to recovery. Others would have to take over from here.

I left Pablo and his friends and headed back to the apartment, having agreed to bring Ghost and all his stuff over the following day. We'd have one more night together in the flat with Nala, and we'd all get to say our goodbyes. It wasn't the easiest of nights. I'd taken a real shine to the little fellow. He had a sweet nature and was beginning to come out of himself. I'd bought some new rope toys and he loved tearing around the flat chasing them, then playing tug-of-war as I tried to take them back. Nala had grown even closer to him and was still sleeping downstairs next to Ghost's bed.

We'd both miss him.

Fortunately, I'd been through this before with Balou. Early on I'd felt rubbish about separating, but it had all worked out for the best. No one could guess that the normal, healthy dog they now saw running around parks in London had been such a pitiful creature at the start of the year.

'You'll be walking the beach in Dunbar next time I see you,' I told Ghost, as he lay next to me and Nala on the sofa.

Knowing that his future was settled made all the difference the following morning when I left him with Pablo. It didn't stop me shedding a small tear, but it did stop me from totally embarrassing myself.

The cycle back through Georgia and across Turkey to Istanbul pushed me to the limits, in more ways than one. For a start, I had the worst border crossing yet leaving Georgia. Until now, no one I'd encountered had bothered to look at the individual stamps in Nala's passport, but the guy on the Turkish side went through each of them one by one, asking me all sorts of questions. I knew everything was in order, but he kept prodding away.

He also asked me to open up my panniers so he could check inside. It was a very quiet crossing point; I think he had nothing better to do. But it reminded me again why it was so important to have up-to-date certification from a vet at all times, otherwise Nala would be in real danger. That particular guard would have confiscated her from me without thinking twice, I felt sure.

I then cycled into the coldest weather I'd encountered so far. I thought I was going to freeze to death on a couple of nights, when we camped out in the low mountains on the road down to the city of Kars.

I booked us seats on another train, this time from Kars to Ankara, a 29-hour journey through central Turkey. But once again someone in a uniform had a problem with Nala boarding. It took one of his colleagues pulling up their own website, and showing him the terms and conditions that said it was perfectly acceptable for a cat to travel on board, for him to let us on the train. At times like these, I was grateful that I'd be on my bike once I got myself back on track, and in charge of my own destiny again. Officials and their rulebooks were beginning to get under my skin.

I'd decided to head to Ankara for a few reasons. It was Turkey's capital and I'd heard it was worth a visit, but I also hoped to start making arrangements for our flight to India, which I knew was going to be a bureaucratic nightmare. Heaven knows how much red tape I'd have to wade through, but it felt like my very best option. The plan was to fly to Delhi, or perhaps Mumbai, then cycle up towards the Himalayas and around.

A part of me was really excited. I'd always seen myself spending a year exploring India; there was more than enough to see. I'd also been contacted by members of one of the charities I'd supported, People for Animals. The head of the charity, Maneka Gandhi, was part of the famous political family. They were talking about sending medical help when we arrived in India. It was going to be a big shock for both of us, but for Nala especially. The vet would help us make the transition.

From India, my plan was to visit Cambodia, Vietnam and Thailand, and then – with luck – cycle down through Malaysia to Singapore and somehow get ourselves over to Australia. I'd already had an invitation to visit Sydney from someone who would help me with their strict animal quarantine rules. He reckoned that, with some planning and advance testing, he could get Nala into the country without having to be isolated for months. The prospect of cycling across the outback and the Gold Coast definitely appealed to me. After hitting a dead end in Azerbaijan, I wanted the world to be opened up to us again.

First things first, however, I had to get all the documentation I needed to get into India. The visa situation was complicated – for both of us.

I'd attempted to talk to the British embassy in Ankara and had cycled over there one day, only to find that I needed to make an appointment. When I went back, I was told to go to the Indian embassy, but the official I spoke to there looked completely flum-moxed. He'd never heard of anyone trying to fly a cat into his country. He directed me straight back to the British embassy.

I'd been ready to give up when a couple of members of Team Nala came to my aid and started making some of the enquiries. After a couple of days it seemed the paperwork was moving in the right direction. Thanks to a Turkish tour operator, I'd been lined up with some provisional flights for the first week of December, but I needed to be reassured again that Nala would be able to fly in the cabin. I'd been clear that was a make-or-break issue.

All that remained was for Nala to get her final checks done by a vet, which I'd get done in Istanbul. India had its own require-ments, for me as well as Nala, but hopefully she wouldn't need anything more than a couple of extra jabs before the flight.

All this admin work was doing my head in, so I was grateful for something more straightforward to occupy me. As November drew to an end, I received a note that the calendars were ready. The designer Kat had done a brilliant job and we'd also got a great distribution deal with an online pet shop called Supakit.

Our first print run was an ambitious four thousand copies, which seemed a lot to me. So when we finally made the announce-ment that they were for sale, I felt a mixture of excitement and apprehension. Would anyone want them, or was I going to end up sitting on boxes of out-of-date calendars?

I got my answer within hours. They all sold out. Just like that. People were ordering them from around the world, sometimes in bulk. The demand caught us off guard, so we quickly organised a reprint, repeating the first run, which was as much as I could afford up front.

I was overjoyed. I did a rough calculation of what would be left after printing and distribution costs and worked out that I could have eighty thousand pounds or more left for charity. Incredible. I was already at work on the list of charities that were going to benefit. It was so exciting to think that I'd be able to help a whole array of causes. It gave me more confidence that I was doing the right thing and a fresh boost of energy to tackle the nightmare of getting to India.

My next challenge was to work out how to transport the bike and my paraphernalia on a plane. The solution seemed to be loading it as special luggage, breaking the bike down and packing it up in a cargo box with my trailer, panniers and all my gear. Then all I'd need to take on the plane would be a rucksack and Nala in her carrier.

As the end of November drew near, I made an appointment to meet a guy in Istanbul who specialised in packing unusual or fragile cargo. Everything seemed to be slotting into place. I left Ankara heading for Istanbul feeling that things were back on track. After so many false starts and wrong turns, the next leg of the trip was happening. Barring any major setbacks or complications, we'd be in India for Christmas.

20
Local Hero

I was halfway to Istanbul, cycling along a busy highway near the town of Sakarya, when I noticed Pablo's name flashing up on my phone.

It was rare that he contacted me, so I pulled over and opened the text message, somehow sensing that it might be bad news. The first few words of the text message confirmed my worst fears.

'Dean, I have very sad news. The puppies are sick.'

I called him immediately. He sounded upset, close to tears.

'The tests have come back and all three have parvovirus,' he told me. 'It's contagious and really dangerous for puppies. Hippie might have it, too.'

I'd heard of parvovirus, a really nasty, potentially fatal infection that affects the intestines, causing severe diarrhoea and vomiting.

'But surely there's a treatment?' I asked him.

'No, there's no cure with medication, for young dogs at least. You have to hope the dog's defences can deal with it,' he said. 'But if a puppy is weak . . .' He didn't need to finish the sentence.

I felt helpless – and guilty. I should have been there with Ghost. I muttered something about money, telling him that I'd raise whatever it took.

'Thank you, but I don't think it will make a difference,' he said. 'We will try our best, of course.'

We ended the conversation promising to stay in touch. I got back on the bike and pressed on towards Istanbul, but my concentration was shot. All I could think about was Ghost. How had he contracted the illness? Had he picked it up during that week we'd spent together? Had he passed it on to the other dogs?

Or had he had the virus for weeks? Did he have it when I first picked him up? These were all impossible questions to answer.

Then I was hit with an even worse thought. *What about Nala?* Was she infected too? Did I need to get her tested?

I booked into a small hotel that night, sitting up until late, talking not only to Pablo, but also to Sheme and others who had helped me before. There was one silver lining. Sheme told me the chances of Nala contracting parvovirus were slim to non-existent. It was predominantly in dogs, not cats.

It was a small consolation, especially as he'd explained how extremely dangerous it is. He'd not sugar-coated it at all.

'I'm sorry, Dean, but I wouldn't put the puppy's chances as much better than fifty-fifty. And that is being optimistic,' he said.

I put a note on Instagram, more in desperation than hope. Our followers immediately sent me links to assorted websites. Some even cited cases where puppies had recovered from parvovirus. It was well-meaning, but none of it could change anything. It was my fault for asking; I was clutching at straws really. No one was going to be able to intervene. Ghost's fate was out of our hands.

I hit the outskirts of Istanbul a couple of days later. It was late afternoon and the traffic had already started to get crazily busy. Cars, buses and trucks were whizzing by me at high speed. I was struggling to follow my sat nav and made a wrong turn at one crucial junction. I didn't feel safe, especially in my current state of mind, so I found a cab to take me to the apartment I'd booked online. It was for the best. I didn't want to get myself – and Nala – killed.

I despaired. I'd been looking forward to exploring Istanbul, and

I had a lot of invites to visit people and places within the city waiting for me there, too. But as I walked the picturesque streets and looked up at its impressive mosques and palaces, I couldn't shake off my mood. I felt torn in two, or three to be accurate. Part of me wanted to dash back to Tbilisi to look after Ghost, and another part was so shaken by the news of the virus that I wanted to stay here and take extra good care of Nala.

As it turned out, I couldn't be with either of them for the next seventy-two hours. The timing couldn't have been worse. I'd made another important promise I couldn't break.

'Would passengers please return their seats to the upright position and fasten their seatbelts for landing.'

The announcement snapped me out of a deep sleep. I buckled up my belt and looked out of the plane's window through the bank of clouds beneath me, trying to make out the grey, urban landscape of London, several thousand feet below. I felt excited but a little nervous, too. I had a busy couple of days ahead. I'd made the decision to dash back to the UK for a very special occasion. My grandmother's ninetieth birthday.

I'd left Nala in Istanbul with two Turkish ladies whom I'd got to know earlier in my travels in the country. Goksu originally made contact via Instagram, but then met me for a drink with her sister Ecenaz when I was passing through her hometown of Antalya. Both of them hit it off with Nala instantly and we'd remained in touch. When I mentioned that I needed a cat sitter in Istanbul, they jumped at the chance and flew up to spend a few days with her at the flat I'd rented. They were experienced cat sitters, so I knew she was in safe hands for the short time I was away.

I was in such a rush when I headed to the airport, I'd barely brought any clothes with me, so I probably stood out as I took the train back up north via Edinburgh to Dunbar. There weren't too many guys in shorts and vests at that time of the year. When I got to Dunbar shortly before 11 p.m., I headed straight for a pub in town. I'd arranged to meet my old travelling companion Ricky.

We'd barely communicated since we split up, almost exactly a year ago. So I was delighted when he replied to the message I'd sent him, asking if he was around while I was back in Dunbar. I often worried whether he held a grudge. I wouldn't have blamed him. I'd begun to fulfil the ambition we both shared when we'd set off the previous year. By contrast, he had apparently slipped back into the same old routine in Dunbar. He had every right to feel cheesed off with me. But the minute I walked into the pub, he stood up and gave me a giant bear hug.

'Hey, look who it is, the local hero,' he said. 'Can I get your autograph?'

I told him to take a hike, but in more colourful language, naturally. I then grabbed us both a beer.

It was as if we'd never been separated. We were soon joking and joshing with each other. It was a huge relief.

He also had a couple of surprises up his sleeve.

First of all, he revealed that he'd stayed on in Mostar the night we separated. I thought he'd been many miles away, but in fact we'd spent the night only a couple of streets apart from each other. We could have thrashed it out over a drink that night if I'd known. His other news was that he had fresh plans to travel. He was going to head to Australia in the New Year and see where his journey took him from there.

'Maybe I'll rescue a koala or a kangaroo along the way,' he laughed.

'As if one would want to hitch a ride with you,' I replied.

Deep down, though, I was thrilled for him.

'You'll no regret it,' I said, raising my glass to him. 'It's changed me as a person. Made me appreciate life so much more.'

We left the pub and found a spot in a garage near the seafront where we sat chatting late into the night and the small hours of the morning. I felt quietly elated afterwards, from clearing the air with Ricky and hearing that he bore no hard feelings. It wasn't long before I was sent crashing back to earth.

I'd kept my trip secret from everyone but my sister. I wanted to surprise my mum, dad and gran. So I'd arranged to stay with Holly and her partner Stuart nearby. They'd moved into a new home in the town a five-minute walk from my mum and dad's house on the seafront.

In the morning, they were keen to hear about Ghost. They loved looking at my latest photos of him, but I had to break the news to them about the parvovirus.

It was soon afterwards that Pablo's name flashed up on my phone.

I'd sent him a couple of texts the night before, wondering how Ghost and the other dogs were getting on. He'd been a bit cryptic in one of them. It seemed Hippie and the puppies were doing okay, but he'd not said anything about Ghost.

It was soon clear why. This time I had only to read the first word of the message: 'Sorry.'

I'd told myself to be braced for bad news, but it still hit me like a thunderbolt. I dropped the phone, devastated. Holly had popped to the kitchen, but came running back in asking what was wrong. I could barely get the words out.

'Ghost didn't make it,' I said.

She was almost as upset as me. All those plans she and Stuart had made would be for nothing. They were both gutted.

When you suffer a loss, it's too easy to blame yourself. It's human nature, I guess. More often than not, there's nothing you could have done. But I couldn't stop thinking: what if I'd stayed with him rather than going into Azerbaijan? What if I'd stayed in Tbilisi through the winter with the three lads? Maybe I could have nursed Ghost back to health, then found him a home in the spring when I set off for Turkmenistan?

The truth, of course, was there was little that I could have done. This virus was a killer. It has a long incubation period, apparently, and Ghost could have had it before we even met. Nothing consoled me, though. Even the words of support I got from my followers on Instagram when I shared the news didn't ease the pain. If anything, it made me feel more guilty, more as if I'd failed Ghost.

I knew I had to put this news to the back of my mind; there was nothing more I could do and I had to focus on my gran and her party. But deep down I could already sense a change in me, that I'd reached another turning point in my trip.

My grandparents – my mum's parents Agnes and Bill – were a huge part of my youth. My old grammar school backed on to their house in Dunbar and I'd park my bike there each day before heading over the wall in time for morning assembly. I'd often pop in for my lunch or some dinner if my parents were working late, as was often the case. My gran's macaroni cheese was legendary. No one had done more to raise me and set me on my way in life; I looked on her as another parent.

Her big party had been arranged for the following day, Saturday, in the 'Mac', one of the main hotels and bars in Dunbar. Relatives were coming from all over Scotland, as well as from my dad's side of the family in Newcastle. It was going to be a big affair.

As much as I would have loved to surprise my gran on the day, I didn't want to give her a heart attack. So, with my sister's help, I headed over to the family home that afternoon.

Still, I'd not been able to resist having a bit of fun.

I've been famous for my practical jokes since I was a kid. I once shaved my dad's right eyebrow off, while he was sleeping on holiday in Fuerteventura. The first he knew of it was when he walked past a bar where I'd written a notice on the chalkboard: '20 euro reward for Neil Nicholson's missing eyebrow'. My family turned the tables on me a couple of times, too. Once I pretended to have found a tooth in my gran's macaroni cheese – but then my mum arranged for me to be sent a fake compensation letter and vouchers from the macaroni manufacturer. I fell for it hook, line and sinker.

So, after surprising my mum and dad by walking into the kitchen at home, we cooked up a plan.

My gran lived in the same house, but had been out having her hair done. When she came back in, my sister and dad told her my mum had taken to her bed because she was finding the stress of the party too much.

They knew my gran would come up and check on her. But when she came into her bedroom and pulled down the duvet – guess who was lying there. My mum was a little worried that it would be too much of a shock, but her reaction was priceless. She was over the moon and gave me the biggest kiss. She didn't stop cuddling me for the rest of the day.

It was great catching up with the family gossip while getting the celebrations for my gran's ninetieth under way. We were soon swapping stories, a lot of them about what a tearaway I'd been as a child.

My gran couldn't resist recounting the story of how, when I was nine or so, I flew over the handlebars of my bike riding along the seafront in Dunbar, fracturing both my wrists in the process.

'We didn't think you'd be going anywhere much on a bike after that,' she laughed.

My mum remembered, when I was around the same age, how I'd been grounded for some reason but secretly sneaked out of the house to go to a youth club disco.

'You hid a bag of extra clothes in the garden hut and climbed out of the kitchen window,' she smiled. 'You always would go the extra mile for a party.'

At dinner time, we started to talk more about my travels, and I explained how hard I'd been finding it to make my way through Central Asia to the Far East and India. My mum and dad seemed very relaxed about it, probably more than me, truth be told.

'I know you're a survivor. You'll maybe not always make the right decisions, but you'll make decisions and that's the main thing,' my dad said. 'I'm just massively jealous of you. I'd love to go for it like you've done.'

It was my mum's comment that surprised me the most, though,

when she told me, 'I worry less about you when you are away than I do when you are home.'

The party the following day was packed to the rafters. There must have been a hundred people there. It was great to see some of my extended family and I spent the first hour or so circulating the room, saying hello to cousins, aunties and uncles. They all seemed to know about my adventures.

'Look what the cat's dragged in,' one of my cousins joked.

'Except I left the cat behind,' I said.

'Shame, we were looking forward to meeting her more than you,' he shot back.

I was used to the teasing and it made me feel totally at home, like I was back where I belonged. Everything seemed the way it had always been. Well, in all ways but one.

Before I left home, few people ever expressed much interest in what I did at work, or what I was up to with my life. They knew the answer to both was usually the same. *Not much*. They rarely asked my opinions on serious subjects, either. I wasn't a serious person; I was a party animal and a free spirit in most people's eyes.

It was very different today. People couldn't stop quizzing me about where I'd been, where I was headed next, what my plans were. Everyone seemed to see me in a new light. I wasn't the village idiot anymore. If anything, people were treating me like a kind of role model.

One friend told me that, like Holly and Stuart, he and his girlfriend were thinking of rescuing a dog from an overseas sanctuary. He'd seen the news about Ghost on Instagram earlier that day and was sympathetic.

'But don't give up,' he said. 'What you're doing is a really good thing. So keep going.'

I didn't show it, but his comment meant the world to me. Thousands of strangers said positive things on Instagram and YouTube each week. But to get the same encouragement from people who'd known me since I was a toddler was another matter entirely. To hear those closest to me, those who knew me best, tell me that they admired what I was doing meant a hell of a lot. Probably for the first time in my life, I felt proud of myself. Not that I could tell anyone, of course. Dunbar isn't a place where anyone is allowed to get too big for their boots.

The party went on late into the night, but as I headed home along the seashore, the familiar coastline visible in the moonlight, it was great to smell the Scottish sea air, to feel the stiff North Sea wind on my face again. It wasn't nostalgia or homesickness. Quite the opposite. I was already itching to get going again. I loved home, but now that I'd popped back and seen everyone, I was ready to head off again.

Everyone was well and happy. Everything was ticking along the way it always did. It was good to know it was there for me when the journey was over, but that was a long way off.

As far as I was concerned, my place now was with Nala. I wanted to continue my journey with her, to finish off what we had started.

I set off back for Glasgow airport the following day revitalised, but also knowing that it was time to make some big decisions.

My mind had been working overtime since the news about Ghost, but with so much going on in Dunbar, I'd found it hard to see the wood for the trees. The flight back to Istanbul was exactly what I needed. It gave me a few hours to mull over everything, to take stock of where I stood. I had a good idea of where that was.

I'd had a couple of conversations back in Istanbul before heading to Scotland that had unsettled me. The first was with another blogger and YouTuber, who was travelling across India on his motorcycle with a cat. We followed each other and chatted privately. He told me that the roads in India were unlike anything he'd seen. Cars, motorbikes, trucks and tuk-tuks whizzed around, oblivious to the road markings and laws. He'd almost been nudged off his bike a couple of times.

A day or two later, I'd had another chat, this time with one of the vets who was going to give Nala her jabs. He'd expressed concern about Nala getting sick.

'The change of climate and the way of life will be a big shock to her system,' he said. 'Humans have trouble when they go there and it's the same for animals, so you will have to watch her health really closely.'

His warning took me by surprise and the shock on my face must have been clearly visible.

'Are you sure it's the right move for you?' the vet asked me. 'Can you not get to the other side of the world by bike? That way you and Nala will adjust to the different climate more steadily.'

As I sat on the plane that night, central Europe lit up beneath me, I began to see that they both had a point. India was going to be a big challenge for Nala – and for me. We were going to make a huge jump into the unknown, and maybe it was one that we didn't have to take.

With these misgivings, I took a long look at the maps and also read a few blogs and web pages by around-the-world cyclists. There was definitely an alternative route. I could start cycling north into eastern Europe and from there head into Russia, crossing it via Moscow to the Far East. At the farthest eastern city, Vladivostok, there were boats to Japan and Korea. I would go the long way around, across country, by land rather than air.

It was a long way. Russia is the world's biggest country and extends a staggering six thousand miles from west to east. It was cold, too, with incredibly harsh winters, which might test even a hardy Scot like me. But that didn't matter; I wasn't on a deadline. I had all the time in the world. My priorities lay elsewhere.

By the time the 'fasten seatbelts' sign was illuminated again and the plane started its descent into Istanbul, my mind was made up. Nala's welfare was too important to take the risk of India. I'd lost Ghost. There was no way on earth I was going to lose her.

21

One Man and his Cat

It was the week before Christmas, but I wouldn't have known it. There wasn't a fir tree or fairy light in sight, not a carol singer or church bell to be heard. For the third day running, Nala and I were marooned in a muddy field in the south of Bulgaria. Fogbound.

The mist was cold and clammy and so thick you could almost cut into it with a knife. On the rare occasions I stuck my head out of the tent, I could barely see ten feet in front of me. All I could hear was the occasional passing car or lorry on the road a few hundred yards away. Even the birds had fallen quiet. At another time I might have found it a bit eerie, scary even. But at this precise moment, the thought that Nala and I were completely alone in the world wasn't so bad at all. It felt like a welcome break.

It was now a week since we'd left Istanbul. We'd stayed in the city long enough to celebrate our first anniversary on 10th December. It was amazing to think it was twelve months since that Sunday morning on the Bosnian mountaintop. We'd been through so much, seen so many places, met so many people. We'd both grown up in our very different ways, too. I was certainly a year older and wiser. I felt as if I'd been through a crash course at the university of life.

We had also stayed on in Istanbul long enough to see the final batch of calendars go on sale. We'd printed another 8,000 and they'd sold out almost immediately. When I tallied up the profit, it was amazing. I now had roughly £90,000 to donate to charities. I didn't want to sit on that kind of money for long, especially when it could be doing so much good elsewhere, so I was hard at work on the long list of causes to which I'd be contributing over the coming weeks. My plan was to give thirty smaller, less well-known charities £3,000 each. I hoped it would make a real difference to them.

All in all, I had a new, positive feeling about the trip and our future plans.

Making the decision to cancel the flight to India was now a relief. I felt a weight lifting off my shoulders. With the complications of flying halfway across the world forgotten, I was free to focus on getting back to doing what I enjoyed most: hitting the open road with Nala.

Being back on the bike had been a rude awakening. Heading north out of Istanbul, I clocked up fifty miles on the first day and I felt every yard of it when I pitched up that night. Not to put too fine a point on it, my backside hurt. Travelling on trains had softened me up, I realised, as I lay in my hammock trying to ease the soreness.

I was also reminded of how unpredictable the weather can be in this part of the world. I'd been looking forward to sleeping rough again, making do in disused or deserted old buildings that I found along the way. I'd found a couple of derelict sites already. But on our third night, within reach of the Bulgarian border, we slept out under the stars. My weather app on the phone predicted a clear night and, with winter coming, we wouldn't get many more opportunities. It was a big mistake. A couple of hours into the night, I was caught by a sudden downpour and had to take refuge in a small hotel. When would I learn to stop trusting weather apps?

Leaving Turkey to cross into Bulgaria wasn't the smoothest ride, either. The Turkish border guards told me I'd overstayed my visa by three days.

As a result, I either had to pay a fine or face being barred from re-entering Turkey for five years. I paid the £30 fine. Given the way my trip had gone, I couldn't take the risk. I might have found myself back there again within a week. I wasn't too angry with the officials: it was my fault for not checking the date on the visa.

My sagging spirits weren't helped when the weather started deteriorating. As we made our way into Bulgaria, the sky became an ominous canopy of dark grey. The light was so poor and gloomy, it was as if someone had hit a dimmer switch. It was hard to believe there was a sun up there somewhere. And then the fog came. For a while it was clear enough to keep cycling, but by the time we passed one of the first sizeable towns, Svilengrad, conditions had worsened dramatically. One minute I could see a hundred or so yards ahead, the next I couldn't see more than twenty feet. If that.

It quickly became dangerous. Cars, vans and trucks would appear out of nowhere and one or two had to veer wildly to avoid us. The landscape was hilly and the roads twisty. I was worried that we'd go over the brow of a hill, or turn a sharp bend, and cycle straight into an oncoming vehicle.

So I'd taken the sensible option and found a spot in a field, a safe distance from the main road, sheltered by some thick bushes. The weather app was forecasting the fog would last for another day at most, but I began to take its predictions with a pinch of salt. It might be over the next morning, it might take a week.

I set up the tent, emptied all our supplies into it and settled down for the long haul with Nala. I figured I had enough food to last us a few days. In a lot of ways this took me back to happier, simpler times. It reminded me of our first week together, almost exactly a year earlier, back in Budva. At that point it was the two of us against the world. No one knew we were together. No one cared. We were one man and his cat defying the elements and living under canvas. I liked it that way. It reminded me what a joy it could be simply being alone with Nala.

With the drama around Ghost and my trip back to Scotland, I felt that I'd neglected her these past few weeks. It was good now to have her to myself, to focus on her full time. She certainly couldn't complain about being ignored during our first day inside the tent. I've never spent so much time tickling and wrestling with her. That first evening under canvas she spent so long chasing the toy I was dangling on the end of some string, she passed out with exhaustion after she'd had her evening meal.

While she snored away, I lay in the tent chilling and listening to music. Thankfully my batteries were fully charged, so I was able to use my laptop for emails and to watch the odd video on YouTube. I was careful to conserve my energy, though. I hadn't forgotten the panic that night when we were visited by the bear in Turkey. If we needed to move, I wanted some functioning lights this time.

I woke up after the first night hoping the fog might have lifted enough for us to get down the road to the next town, about five miles away. But if anything, it was hanging there more stubbornly than ever. I let Nala out to do her business and went for a stretch myself, but didn't wander more than a few yards to the bushes nearby. It was dark by late afternoon, so any hopes of getting back on the road were soon dashed. It was the same story the following day.

As we settled down for our third night, I was beginning to worry about food supplies. Nala's stock of snacks was fine, but I was down to a few cans of beans and some coconut water. Another night and I'd be in trouble, but I wasn't panicking. I'd faced bigger tests. If I had to go hungry for a day, then so be it.

At last the fog in my mind began to lift. I started using the downtime to put some flesh on the bones of my new travel plan. It had one guiding principle. I was determined not to make Nala's life any more stressful than was necessary, which meant no planes for now. Further down the road in Russia, we could always use the famous Trans-Siberian Express to hop between major cities when we needed; Nala was very happy on trains. But even then, I hoped to be cycling through much of the warmer Russian spring and summer months.

Of course, long-term, it was going to be impossible to circle the world entirely on land. But if we got to the farthest eastern part of Russia in a year or so's time, I knew we could get a boat to Japan or Korea. Nala would be two years old by then and would have acclimatised to being in a different part of the world. I wanted to avoid any big shock to her system. The memory of Ghost was too fresh.

When I stuck my head out of the tent on the fourth day, the visibility had improved enough for me to make a dash for the next town. I hoped to get to Bulgaria's second largest city, Plovdiv, by Christmas Eve and it was now 20th December. With a little luck, I could make it.

The light was still poor and the traffic prone to getting too close, so at the first petrol station, I treated myself to an early Christmas present – a high-visibility yellow jacket. It was never more needed.

I landed in Plovdiv on Christmas Eve. I'd rented a nice apartment and decided to stay put for a couple of weeks. It was a European Capital of Culture, so there was going to be lots to enjoy.

Everyone else took it easy over Christmas, why shouldn't I?

After wishing my online followers a happy Christmas, I let them know I was shutting up shop for the holiday season. I'd post only the odd photo on Instagram and I'd get back to the YouTube channel in the New Year. Until then, I was going to recharge my batteries and find my bearings again. Everyone understood, of course. All I got back was warmth and good wishes.

Christmas Day was a quiet affair. Nala and I shared some decent food I found in a supermarket nearby, then I relaxed watching movies and videos on YouTube, but not before speaking to my mum and dad and family back home. Seeing them did make me feel a little homesick. I'd found the last Christmas difficult, but this year it was even harder being away from home. I put it down to the rollercoaster of emotions I'd been through these past few weeks. I was burnt out, mentally more than physically.

Once again Nala was my teacher. She had really taken to the flat, especially the little balcony overlooking the pretty street below. She could spend hours there watching the world go by, snatching a snooze when she felt like it. I needed to take a leaf from her book again; I needed to switch off and enjoy life day by day. I did exactly that.

It was a few days after Christmas that a call flashed up on my phone. I couldn't believe the name on the display. It was Tony, the kayak guide from Santorini.

'Hey, Tony,' I said, pleased to hear from him.

'Hey, Dean. Guess where I am?'

I vaguely remembered something he'd said about having studied in Bulgaria. He'd also said something about having a flat there, but I'd no idea it was in Plovdiv. What were the chances of that?

He explained that he'd travelled from his home in Athens to collect a car he'd bought from someone in Plovdiv, and was staying in the city until the New Year. He'd been following me on Instagram and spotted that I was here too.

He had some news. I'd met his girlfriend Lia a couple of times on Santorini. Tony announced that they'd got married a few weeks earlier. Not only that, but Lia was pregnant.

'We'd better celebrate,' he said.

So my calm and quiet existence was soon turned on its head. In the run-up to New Year, it was like Santorini all over again. We partied for what was left of the old year, raising a glass or five to the newlyweds and their soon-to-be parenthood. To be honest, it did me the power of good and I appreciated the company.

I spent New Year's Eve at a great house party thrown by one of Tony's friends. It wasn't far from the flat where I'd left Nala safely tucked up with lots of treats.

'So, what's your New Year's resolution?' Tony asked me, a mischievous smile on his face. 'Go round in circles for another year?'

'At least going around Europe and Asia beats going round in circles in a kayak on the same island every day.'

'Very funny,' he laughed.

I explained my plan to him.

'If I can get into Russia, the whole world will open up to me,' I said. 'If we're lucky, we'll be in Japan by late spring, then Thailand and Vietnam by the following summer.'

'Be careful if you do go there,' he said. 'Russian roads are really dangerous.'

He wasn't the first person to mention that, but I knew now that there was no route without its risks, so it didn't faze me.

By the first week of the New Year, Tony had returned to Greece and I'd started to make preparations for the next leg. We'd seen some serious snow in Plovdiv over the Christmas break. I knew there would be plenty more of that ahead of us in eastern Europe and my biggest priority was to make sure Nala and I were well insulated.

Nala's basket had served us royally during the summer months, but it wasn't going to be warm enough now, so I invested in a much more solid and well-insulated carrier. It was bucket-shaped and waterproof with a detachable roof as well. It was perfect in all but one detail – the logo on the front was of a dog. Fortunately, Nala forgave me for it and didn't complain. I also ordered another set of tyres from Schwalbe – special snow and ice ones with small spikes for extra grip – as well as two completely new wheels.

I reckoned I was going to need them soon.

I started planning my trip through Russia in more detail. Midway through January, I'd had a conversation with the Russian tourist board based at the Russian embassy in London. I'd also spoken to a specialist company who organised travel in Russia and beyond. They'd given me another option for crossing Russia, which really captured my imagination. It meant travelling by bike and train into Siberia and catching a train from there through China to Saigon in Vietnam. I'd always intended to cycle through Southeast Asia, but I knew going to China would involve a lot of extra red tape and Nala would face more injections. This way would be a lot easier, because we wouldn't be getting off the train. It sounded like a great solution to opening up a whole new part of the world for us.

Neither route was going to work, however, if I didn't get a Russian visa first. It couldn't be an ordinary tourist one, either. They lasted only thirty days and I had 6,000 miles to cycle, even if I did hop on a few trains. I also fancied the flexibility to detour into other countries such as Kazakhstan and Mongolia along the way. I even wondered about dropping down into Uzbekistan, so that I could pick up a part of the Pamir Highway there.

Both the embassy and the tourist agency suggested I try to get a business visa for a year. It required a letter of recommendation from the Russian government, but the guy I was speaking to at the embassy in London, Victor, was confident he could get it for me. I sent him a lot of information about us, from the newspaper articles to my Instagram and YouTube accounts. Between the two, I now had more than 800,000 followers. I also explained that my intention was to show beautiful parts of Russia that most people would never see. I had no idea who would make the final decision, maybe it went all the way to the Kremlin; but I'd given it my best shot.

As January drew to a close, I felt more positive about not taking the flight to India. Our new path ticked all the boxes – for the trip, for me and, most of all, for Nala. Suddenly I could see a way across the world. I felt excited and I'd got my mojo back. I left Plovdiv and hit the road again on the last day of January.

The early February weather was a shock, even for a Scot. I woke up one morning, after a night out in the tent, to find that my groundsheet had frozen. My face was pressed against ice.

When I looked out, I saw the ground was covered with heavy overnight snow. For Nala, it was the best news ever: the world had turned into a giant playground. She'd been out in snow before, in Plovdiv over the New Year holidays, but it was as if she was discovering it for the first time. It was hilarious to watch her gingerly putting her paw into the white powder. As she took her first tentative steps, she glanced anxiously at me, looking for guidance. *What's this weird stuff? Brrr. It's really cold!*

Her caution was soon cast to the wind. She spent the next ten minutes rolling around in the snow, dipping her head into it every now and again, then stepping back to admire the hole she'd made. At one point, while she was skipping and dancing her way around the edge of a large drift, I was unable to resist throwing a little snowball at her. It was worth it for the look on her face as it whizzed past her ear and *thwumped* into the snow behind her. I don't know if it was shock or excitement.

The brief narrowing of her eyes and the tilting of her head that followed was much easier to read though.

I'm gonna get you for that, mister.

By the end of the first week of February, I had crossed into Serbia. The roads in this part of the world were in great condition; the asphalt surface and markings looked brand new. There was also lots of space for bikes on the inside lane. It made cycling a real joy and I was soon making my way north at a decent clip.

I reckoned I'd be in Hungary by the end of the month. From there, I'd either go into Poland via Slovakia, or maybe the Czech Republic; the latter looked easier, as it would allow me to follow the River Danube. The roads would be less hilly that way.

On Valentine's Day, we arrived in the town of Nis, where I'd been invited to stay by a couple of followers, Katarina and Jovana. They cooked me a wonderful meal and even bought Nala a Valentine's Day present and card.

The kindness of strangers never ceases to amaze me.

From there, I carried on bound for the Hungarian border. All was going to plan, I was travelling faster than I had at almost any time in the journey, clocking up fifty or even sixty miles a day with regularity. I was confident I'd be in Budapest in March, and perhaps even Moscow by June.

The only cloud on the horizon was some news from the tourist guy in London. It was Murphy's Law. No sooner had we come up with the plan to get through China to Vietnam than a new wrinkle appeared. Some kind of flu or virus was causing travel problems in China.

I'd been vaguely aware of it in the news, but hadn't paid too much attention. That is one of the great advantages of being on your own on a bike – you can shut out the rest of the world and its problems. But emails from the travel agent, a guy called Yuiri, began drip-feeding me information about the growing seriousness of the illness. Apparently, it had become massively infectious and dangerous in the city of Wuhan in central China, and was killing people, mainly the immunocompromised and the elderly. He told me eighty people had died there and the whole city was in lock-down before it spread any further.

His most recent email was telling me things had got worse. It had spread outside of Wuhan, reaching Hong Kong. The Chinese government were imposing wider restrictions, especially for travel by foreigners.

'It might be difficult to get the train down through China to Saigon. We are being told that we can't organise any tour parties there,' Yuiri had written.

I was disappointed. It had seemed like the perfect solution. But I knew I'd have to roll with the punches and find us some other options.

It's not as if the whole world is shut down, I thought.

Talk about famous last words . . .

22
Number One Fan

By late February, the snows had all but disappeared and the first, faint whiff of spring was in the air. The crisp, sunny mornings and clear blue skies made cycling really enjoyable. It was also perfect weather for sleeping outdoors. One evening, after passing through a little town called Velika Plana, about fifty miles south of Belgrade, I set up camp deep in some woods. I've always enjoyed sleeping amidst trees. I love the smells, the sounds, the feeling that you are wrapped up under a natural canopy, even when, as now, the trees were mostly bare. It was soothing, I usually slept like a baby. After a quick dinner, Nala and I were soon snoozing away in our hammock.

At about five o'clock in the morning, I was woken by the sound of dogs howling nearby. Nala had registered them too and looked around anxiously for a moment, sniffing the air as if to check out any danger. The howls faded, but we were both unsettled. Neither of us slept well from then on and by the time the sun was up, we were restless and awake.

Then around eight o'clock, I heard a voice somewhere in the woods.

I'd spotted a woodland trail not too far away from our camp, so at first, I thought it must be a dog walker talking to their pet.

But after a few moments, I realised the voice was talking in English. It was also getting closer.

'Hello, hello.'

I couldn't believe the sight that greeted me when I poked my head out of the hammock. A lady was standing about ten feet away. She was young, in her twenties probably, and smartly dressed. She was smiling and holding something out. It was a flask.

'I have made you some coffee,' she said.

She then dipped into her coat pocket and produced a tin.

'And I brought some tuna for Nala.'

I had to do a double-take. Why was a woman delivering me coffee in the middle of a forest first thing in the morning; had I accidentally butt-dialled a Serbian branch of Deliveroo? And how did she know Nala's name? My mind raced for a moment. I'd put a photo up on Instagram the night before of Nala testing out the hammock, but she was surrounded by hundreds of trees, all of which looked the same. How on earth had this lady found us based on that: was she some kind of forest detective or expert tracker? This was too much for my head to take in.

I didn't want to appear rude, so I clambered out of the hammock.

'Ah, thank you, that's very kind,' I said, accepting the flask and pouring some coffee into my metal mug.

'If you would like, you can come to my house for breakfast,' the lady said. 'It is not too far.'

I wasn't going to look a gift horse in the mouth, so I packed up and accepted the offer. She'd parked her car on the road and I followed her on the bike for about five minutes to a small house on a farm. A couple of tractors were parked near some outbuildings, where chickens, ducks, and a few cats were running around. She led me into her house, made me another coffee, and put some food down for Nala.

A couple of cats were wandering around in the house, but after a brief standoff and some hissing, they backed down and skulked off to their favourite spots elsewhere.

'Don't worry, Nala, they won't bother you,' the lady said, stroking her on the back of the neck as she tucked into her food. I'd noticed that Nala had been rubbing up against the lady's legs as she stood at the stove. It was a sure sign that she approved of her.

'I'm sorry, I haven't asked. What's your name?' I said.

'Call me Jovanka,' she replied.

'Nice to meet you. Say hello to Jovanka, Nala.'

She told me that her husband was working in Switzerland. She was employed there too and was flying back to her job the following day.

'That's why you speak such good English, I suppose?'

'Thank you, yes I need to speak English for my work,' she said. 'It's a stroke of luck that I was here when you were passing through. If you would have been here tomorrow, I would have missed you. I would have been so upset. I'm a big fan. I have followed you on Instagram since the beginning.'

She cooked me a delicious breakfast, Serbian style. Eggs and bread and tomatoes.

'So I've got to ask you,' I said, halfway through my meal. 'How on earth did you find me in the woods?'

'The same way that lady found you at the bus stop in Turkey. On Instagram,' she explained.

I was amazed she remembered Arya back in Sivas.

'But that was a bit different,' I said. 'I was at the bus station in a big town. That wasn't hard to find. You found me in the middle of a forest.'

She smiled.

'Ah, well, when I saw your post, it had the name of the town, which is the one nearest to here. I showed it to my husband and he knew straight away where you were.'

'How?'

'He walks in those woods often and he recognised it. I didn't believe him, so I went there at five o'clock in the morning to check.'

'You saw me at five? That's when I heard some dogs howling.'

'Ah, that might have been me, I suppose. I parked my car near a farm where they keep some big wolfhounds.'

I was shaking my head and smiling, and my disbelief must have been obvious. For a moment, Jovanka looked embarrassed.

'Sorry, you must think I am some kind of crazy stalker. But I didn't want to wake you too early, so I came back later.'

I couldn't help it; for a moment I was reminded of that film *Misery*, where a writer stumbles across his 'number one fan', played by Kathy Bates. She turns out to be a psycho who is so obsessed with him that she keeps him captive.

Given the lengths she'd gone to in order to track me down, this could very well be my number one fan. The more we talked, the more Jovanka seemed to know about me and Nala and our adventures together, but I could tell that she was a really sweet and slightly shy person and I didn't need to worry. Nala had given her the thumbs up, too.

Another cute little white cat wandered in, as if smelling the food that Jovanka had put down for Nala. She scooped some food into an extra bowl and placed it at the other end of the kitchen.

'How many cats have you got?' I asked.

'Five, but we see a few strays from the village and local farms as well. They have a lot of space here. And my parents are on the farm here as well. They can look after them when we are away.'

Her cat had eaten her food by now and jumped up next to her at the stove, where she started rubbing her cheek close to Jovanka's face.

She smiled at me.

'Unconditional love. That's the beauty of cats, isn't it? That and the fact there's no judging you or making demands.'

'Well, not too many,' I smiled back. 'You should hear Nala when she wants her breakfast first thing.'

Jovanka laughed.

I found it very easy to talk to her, which was fortunate, because she had all sorts of questions about my adventures. Midway through the morning, we were still talking away.

The weather had turned colder and it looked like there might be rain, so I accepted her offer to stay for lunch. She was cooking pancakes.

'Great,' she said. 'Would you like a drink?'

She held up a large bottle of Hendrick's gin.

'Gin and lemonade?'

'Only if you have one too,' I said.

'Okay, why not?'

Any plans I had to get back on the bike that day went quickly out the window. I accepted her offer of an evening meal and overnight stay. She had a mattress set up in her garage that looked perfect.

As afternoon moved into evening, our conversation got more and more relaxed.

'So, why do you follow us?' I said.

'Her, of course,' Jovanka nodded at Nala, who had draped herself across a chair and was fast asleep. 'I love those videos where she's sitting in the front of the bike, watching the world go by.'

'The Nalacam,' I said.

'Yes. But I like the fact you are so crazy as well. I can't believe some of the places you end up sleeping. And the jokes you make.'

Unlike some of my more sensitive followers, she loved the video I'd made a few months back when I intercut some different clips to make it look as if I'd sent Nala up in the air, attached to my drone.

I sensed an opportunity to answer a question that had been bugging me since I'd left Istanbul.

'Were you disappointed that we didn't go to India?' I asked her.

'Disappointed, why?'

'Well, the page is called one bike one world. I'm supposed to be going around the globe,' I said. 'But Nala and I are almost back where we started.'

It was true. I'd looked it up. As the crow flies, my current location was about two hundred and seventy kilometres from Trebinje, from where I'd set off that fateful Sunday morning back in Bosnia.

She poured me another glass of gin and lemonade.

'As long as you are safe and well, I don't think anyone worries about which route you pick and how long it takes you,' she said. 'People follow you because they care about you and you entertain them as well.'

It was what I hoped was true, and it was reassuring to hear someone else say it. Especially someone who had taken such an interest in us from the start. We sat up until late, sharing a few more gins, talking about all sorts of things.

Looking briefly at the news on my phone at one point, I saw that a cruise liner and its four hundred passengers was now in quarantine off Japan after one of them had the sickness – the coronavirus, as people were now calling it. People were worried it was spreading to Europe.

'Imagine being locked down like that? Would drive me round the bend,' I said. 'Three nights in a tent in the fog in Bulgaria and I was going stir crazy.'

'Sounds like it's pretty serious, though,' Jovanka said. 'I heard they had some cases of it in France and Italy already.'

'Glad I'm heading in the opposite direction then,' I remarked.

Nala and I slept like logs in the garage, but I woke up the next day with a raging hangover. My head felt as if it was being squeezed in a vice. Jovanka, on the other hand, didn't seem to be suffering from the previous night at all. She insisted on cooking me another huge breakfast, even though she was getting packed to catch a plane to Switzerland later that day. She rustled up a load of snacks for us, too, from sandwiches and cakes to a litre of gin, which she'd produced from somewhere.

'Sorry, I don't have room for that,' I said, politely handing back the bottle. 'And even if I did, I don't think I can face another gin again for a while.'

'Ah well, more for me and my husband, I guess,' she laughed.

She gave me her Instagram handle and we promised to stay in touch. I was soon back on the road and reached Belgrade by evening.

It was only that night, as I lay in my hotel room, that I looked up Jovanka online. She hadn't been lying when she told me that she wasn't a fan of social media. Even though she must have been on Instagram for a long time to have followed us so closely, she had now made her very first post – a selfie of her with Nala and me and the bike. She'd written a short but sweet note with it, thanking me for my videos and stories.

'You make a lot of people happy . . . sharing your journey with Queen Nala,' she wrote. 'Wish you a safe trip, your Hendrick's stalker.'

After a few days in Belgrade, I pressed on towards the border with Hungary and crossed over at the beginning of March. I followed the River Danube, which made for more easy cycling as well as some spectacular scenery. I was racking up the miles now.

It took me about a week to get to Budapest. I was smitten by the city immediately. Its architecture is stunning, but it also has a real buzz to it, with streets full of cafes and bars. I decided to spend a few days exploring it properly.

Before I arrived, I had been contacted by a lady called Julia, who worked in the tourist industry. She offered to take me on a guided tour of the city, and I couldn't resist. We struck up an instant friendship, partly because Nala liked her so much. It reminded me of the way she'd been with little Lydia back in Athens. The pair of them were really affectionate with each other, almost from the moment they met. There was an immediate bond there.

I'd only just settled into Budapest when I got a note from a couple of friends I knew from Dunbar, Fraser and Maya. They'd come over for a weekend break. It was great to see them and catch up on the news back in Dunbar, but Fraser seemed more preoccupied with heading over to my hotel to meet Nala.

Inevitably we spoke a little about the coronavirus. I'd heard that people back in the UK were starting to advise against trips like theirs.

'It's just crazy,' Fraser said, as we shared a beer. 'Like the world is going mad.'

He wasn't wrong. I'd seen that thousands of people had contracted the virus in Italy, where the death toll was growing by the day. Its government had introduced strict rules on movement. Several cities were in lockdown, with citizens forbidden from leaving their homes.

The situation in the UK was heading that way too, I learned from Fraser. People had been told to keep well apart and not shake hands – to practice social distancing as it was being called. There was talk that pubs and restaurants were about to be closed, and I could see that it was moving in that direction everywhere. America, Canada, India, Australia. I looked up the latest news on Hungary and it was also talking about locking down the population and closing its borders soon.

Given all that was going on, my thoughts turned to my family back in Scotland. My mum worked with vulnerable and isolated elderly people. From what I'd heard, they were also the people most at risk from the virus. At least my mum and dad and gran were together, I consoled myself.

I was beginning to sense that this could seriously affect my trip when I finally received an email from the guy at the Russian embassy, Victor. The letter of authority from the Russian government had arrived. All I needed was to come to the UK, go through an interview and, hey presto, I'd have a visa for a year. I pushed my worries aside – who knew what was going to happen? This was too good an opportunity; it was my guarantee that I'd get around the world at last. The route to the Far East had opened up to me.

The visa wasn't due to come into effect until early April, but I knew from everything that was going on that I'd better make my trip soon.

I asked Julia whether she'd be willing to watch Nala for a couple of nights. She said she'd be delighted. I trusted her to look after Nala, who would no doubt be spoiled rotten while I was away.

So I made the arrangements and booked a return flight the next day. I hurriedly packed a travel bag, then took Nala over to Julia's flat along with her favourite toys and food. I gave Nala a big cuddle, before heading out the door to catch a taxi to the airport.

'Be a good girl with Julia and I'll see you soon,' I said, stroking her neck and giving her a little kiss.

I really hoped that was true.

23
Russian Roulette

The message over the plane tannoy was familiar, as was the sight of the grey and slightly gloomy-looking British countryside spread out beneath me. But as I fastened my seatbelt, I felt very different. Three months earlier when I'd flown back to the UK, I'd been excited. Today I felt jittery and on edge. I kept looking at my watch.

My plan was to get in and out of London in about thirty-six hours. I'd stay here one night, then fly back to Nala in Budapest the next one. For once I was organised and I'd made an appointment to visit the Russian embassy the following day. I'd flown out of Budapest to make it just in time. I couldn't afford for anything to go wrong.

I spent the afternoon and evening before my appointment making sure I had everything I needed. I got a haircut, and a couple of new passport-style photographs to take along with me.

The atmosphere in London was odd. Riding the Tube early next morning, quite a few people were wearing face-masks, wiping their hands with sanitisers and wet-wipes, and trying as far as they could to keep a safe distance. Some seemed on edge. On the breakfast news, I'd seen that there was talk of a total lockdown across the UK within the next week or so, and apparently there was already panic buying in the supermarkets. It felt like the calm before a storm. What sort of storm, I had no idea.

I hopped off the Tube at Notting Hill Gate, then walked down the Bayswater Road. The main Russian embassy is in a grand, rambling old Victorian building not far from Kensington Palace, but the office I was going to was in a more modern building nearby. I got there nice and early, as it was opening for the day.

The official from the tourist board with whom I'd corresponded, Victor, was there to greet me. He was a friendly young guy, not at all the intimidating sort of figure I'd imagined working there. He showed me the official letter of invitation from the government.

'So, they have approved for you to stay on business for a year,' he smiled. 'With the right to enter and leave the country as well.'

I handed over my passport photos and we went through the application forms I'd already filled in online. I discovered there were still a few pieces of red tape to be sorted, including a section that I hadn't completed in detail. It asked me to list exactly where I'd be travelling. I was concerned by stories I'd heard of how travellers to Russia had to account for their every move, and to keep receipts from restaurants and hotels. But Victor quickly put me at ease on that front.

'I think this will have to be a rough plan, yes?' he smiled.

'Aye. Hopefully I can cycle to Moscow, then maybe travel on the Trans-Siberian Express a little of the way. Then do some cycling in different regions. Maybe visit Kazakhstan and Mongolia,' I said.

He started tapping away at his screen.

'So, I will put down that you will visit the big cities, Moscow, Yekaterinberg, Omsk, Novosibirsk, Irkutsk, Vladivostok.'

'Sounds good,' I said.

'You must visit Lake Baikal in Siberia north of Mongolia, by the way. It is the deepest lake in the world. A great place for cycling,' he advised.

'So I've heard,' I nodded, feeling more confident now that I'd be okay. I kept looking at my watch. I had hours yet to get back to Gatwick, from where my return flight was leaving. I'd be back with Nala by late tonight.

'So, can I take your passport?' he asked.

'Sure,' I said, handing it over. I assumed he wanted to check the details that I'd inserted when filling the form in online. I was wrong.

'We will need to hold on to this for a few days to put the visa in,' he said.

'Sorry, what?'

I was totally thrown. I thought it would be done while I waited. Or that the visa would be a separate piece of documentation that I could have collected by someone else, who would then send it on to me.

'A few days? How many?'

'It is possible I could have it ready for late tomorrow night, but I cannot guarantee it,' he responded. 'Most likely in four working days. So next week. After the weekend.'

I didn't want to be rude. He was doing his best, and it was obviously my mistake. I'd not read things correctly or, if I had, I'd misunderstood something.

'Can it not be done quicker?' I asked. 'I need to get back to Budapest.'

'I am sorry,' Victor said. 'We can do it as fast as possible, but we cannot guarantee that it will be ready faster than four days.'

A week or two ago, I wouldn't have worried. Three or four more nights weren't going to make a huge difference and it would have been worth it for the freedom it would give us. But things were changing – fast. Very fast indeed.

It was like I was playing a game of Russian roulette. If I delayed my return, the travel situation in Hungary could change suddenly. The shutters could come down and Nala would be marooned there, while I'd be stuck here in the UK.

We might never see each other again.

Victor was still holding the passport, giving me a look that meant: *The ball's in your court.*

I didn't have time to dither. There were other people waiting to get their visas, each of them – I felt sure – with their own problems.

I had to make up my mind.

'Sorry, I'll have to come back and complete this when everything has died down,' I said. 'I can't take the risk.'

'I understand. The situation is difficult right now. It's your choice,' Victor said, handing me back my passport. 'But the visa is here waiting to be completed and added to your passport whenever you can get back. Good luck.'

It was beyond frustrating. I'd been so close. I'd seen the letter of invitation, they had all the details they needed. But I couldn't take the gamble.

I walked back towards the Tube station cursing myself. How could I have been so stupid? Why didn't I see that coming? I went through the emails that had been pinging back and forth, to satisfy myself there had been no clue that they'd want to hang on to the passport. There wasn't.

Maybe they thought I had a second passport, I wondered to myself. The thought stopped me in my tracks.

Wait a minute – a second passport?

If I could get another passport today, then I could leave it with Victor and it could be sent on to me by post or courier. For a few moments I got really excited. It was still mid-morning, I had time. I made a call to the passport office. I wasn't that far away; it was near Victoria railway station, and I knew that if you paid a premium price you could get a passport in a day. But my hopes were soon dashed. The soonest I could get an appointment was in three days' time. I felt deflated and defeated again. It simply wasn't going to be.

I walked past a newspaper stand and saw headlines about a European lockdown and borders being closed.

What the hell have you done, Dean?

I jumped on the train to Gatwick Airport, a million thoughts flashing through my mind. Should I have hung on for the visa? Was I doing the right thing? Everything was so uncertain. But the more I scanned the news on my phone, the more dire the situation seemed. All I could see were fresh headlines about travel restrictions being imposed. It was as if the whole world was being shut down. I looked for any news about the situation back in Hungary, but no luck. I'd no idea what was happening back there. The border might be closed already and my flight might be cancelled. If it was, I was really screwed.

I got to Gatwick early evening with about two and a half hours to spare. The atmosphere here was really peculiar – you could feel the tension. As I arrived in the departure hall, people were gathered in clusters around the information desks and by the giant departure boards. A couple of businessmen were looking up at the boards, shaking their heads while in animated conversations on their phones. A young Asian girl was being consoled by her partner. It was soon clear why.

I looked up at the board, scanning down the columns looking for my flight. Several had been cancelled. A lot of the others were delayed. As I scrolled down to mine, I held my breath. There was nothing except a note saying: *Wait for further information.*

I headed over to a nearby bar. I needed a drink; it had been a stressful day. Bizarrely, a couple of people recognised me sitting there and asked for selfies. How they'd spotted me without Nala on my shoulder was a mystery to me.

As I sat at the bar, the television above me had rolling news and red banners with the latest coronavirus updates flashing constantly across the screen. More talk about borders closing and flights being stopped. I wouldn't believe I was back in Hungary until I stepped off the plane in Budapest.

The stream of messages I was getting on Instagram didn't help to ease my nerves. I'd put up a note about my dash to London the previous day and people were already panicking on my behalf. A lot of them had worked out that I might struggle to get back to Hungary.

I sent a message to Julia in Budapest. She replied almost instantly. It was obvious she was worried too. On the local news in Hungary that night, there had been talk of the borders being closed sometime in the next day or so. It might even happen tonight, she said.

'What happens if I'm in mid-air?' I asked her. 'Can they turn us around?'

'I don't know. Nobody knows,' she said.

A few moments later, she sent me a photo of Nala, curled up on her sofa. She looked content, as if she didn't have a care in the world, as usual. It was meant to reassure me, I knew, but it did the opposite. I felt a terrible wave of guilt. How could I have abandoned her? How could I have risked everything like this? Would I ever get to see her again?

With less than an hour before my flight was due to take off, my stress levels were rising even higher. I kept dashing back from the bar to the departure board. A lot of the flights that were leaving after mine now had their boarding details showing. But nothing was happening with my flight. I kept willing the entry on the board to change, but at the same time was terrified that it would flip to 'cancelled'. That was happening more and more. The board was peppered with little black notices.

It was a little more than half an hour before my flight's scheduled take-off time when the change finally came. Suddenly there was a gate number. The notice said simply: *Boarding*.

I ran down the empty corridors as fast as I could.

People were being ushered on to the plane already. I headed

down the ramp and on to the plane to discover the cabin was virtually empty. It was me and about a dozen other travellers, all of us spaced out along the length of the aircraft at a new social distance.

The flight was surreal. The stewardesses wore masks and gave us our food and drink wearing gloves. We were offered alcoholic wipes to clean everything on our trays. It was all very unsettling, but I was on my way back.

Now I needed to rethink my options. My instinct was to hit the road. If all the borders were going to be closed, then I'd like to camp in the wild somewhere around Budapest, where I'd be out of people's way. I would be like a character from one of those post-apocalyptic films where there are only a handful of people left on earth. Except in this case, there would be a cat, too. That rather appealed to me.

But when I thought it through, it was much too risky. These weren't ordinary times and it would be easy for a police officer or soldier to decide that I was some kind of risk. I could end up in serious trouble. What I needed was a bolthole, a safe haven for the two of us. I'd have to work something out when I got back.

The plane landed close to midnight local time. I jumped in a cab and went straight to Julia's flat. My heart was pounding as I climbed the stairs and knocked on the door. It was crazy, I'd never been so eager to see someone in my life.

I'd hardly got through the door when Nala exploded out of Julia's arms like a rocket. She clung to me so tightly I could feel her breath on my face. Her ribs felt like they were about to burst out of her stomach, she was breathing so heavily and deeply. If I'd not known better, I'd have thought she understood the situation, and could tell how close a shave I'd had.

That night she curled up alongside me closer than ever, purring in my ear. I've never cease to be amazed at her instincts. We'd been separated before. When I'd gone to see Balou back in Albania, when I'd popped back to see my gran. What was different this time? Maybe she picked up on my anxiety, or she could hear something in the way I was breathing?

Back at the flat the next morning, I scanned the overnight news online. It was even more apocalyptic. The government here in Budapest had announced that it was closing Hungary's borders with immediate effect. They weren't allowing people in or out.

I couldn't believe how close I'd come. I'd made it by the skin of my teeth.

24
A Good Traveller

Over the next few days, the situation was gathering pace; things were changing by the hour.

The Hungarian government began issuing new rules and regulations. People were required to stay indoors and not move around unless for essential business. You were allowed to go to the pharmacy or the supermarkets, but that was about it. Being cooped up in a flat in the centre of the city wasn't great for me, but was particularly bad for Nala. She needed some space to run around. She was still a young cat, after all.

I started looking around online, but couldn't find anywhere suitable for us to stay. Most places were marked as unavailable; the lockdown was well and truly under way.

I toyed briefly with the idea of trying to get back to Scotland with Nala, but had to rule it out as impractical on too many levels. Luckily, the perfect solution presented itself, as I've learned it often will when you stop worrying and let life take its course.

The offer of a home came completely out of the blue, in the form of a message from a lady called Kata, who followed me on Instagram. She and her husband and children had a house in the countryside, about half an hour outside Budapest, but they were going to be stuck in quarantine in the UK. I'd be doing them a favour by keeping an eye on their property.

I messaged her back instantly. It sounded ideal.

Kata told me that her parents lived on the same plot of land, but in a smaller property at the other end of the garden.

I was soon packing up the bike and cycling out of Budapest. There weren't any roadblocks or barriers up at this point, and the police didn't seem to be enforcing the lockdown that rigidly. I made it to the house within a couple of hours. It was down a little lane in a quiet, small neighbourhood on a hill. Spread over three floors, it had all the mod cons I could have wanted. There was even a balcony with a lovely view over the surrounding countryside.

The place suited us down to the ground, although – at first – our neighbours didn't agree.

A couple who lived next door complained to the owners' parents when they saw me arriving. I guess they worried I was an outsider, who might be bringing the virus into their community. I couldn't blame them and I respected their concerns, but it was a shame. I'd have been happy to help the community, delivering food to locals who were vulnerable, or couldn't get out for other reasons. But I realised I'd have to give it time. People would be suspicious of me for now. I stood out in the best of circumstances, and in this situation I would do so even more.

So, a little reluctantly, I hunkered down and tried to give some routine to my life. At least I had Nala to help me.

There was a big garden, which was great for her to run around. I'd cycle the three miles or so to the nearest supermarket, or for some exercise to the nearby lake. But I spent most of the time indoors, talking to people back home, and playing with Nala. I also started playing chess with my dad back in Scotland.

Each time I ventured out, I found the world getting quieter and quieter. The few people I saw looked anxious, in a hurry. No one was talking.

My birthday came at the end of March. It was an odd occasion.

I received lots of messages and spoke to my mum and dad and sister back in Scotland, but the only real person I saw was Kata's father next door, busy in his garden. He acknowledged me with a brief wave, but that was it.

There was a silver lining, of course. There always is. While our world was shrinking, so too was the space between me and Nala.

It didn't seem possible, but we became closer than ever. I spent hours playing with her, either in the garden or around the house. She loved patrolling the patch of grass outside our front door, or tearing up the wooden stairs in the hallway, while I tried to grab her through the gaps. She never got bored, which was something I feared would happen to me if the isolation went on for too long.

At least I had plenty to keep myself occupied in the short term.

For months I had been meaning to go through my archive of photographs. I had hundreds, if not thousands of them stashed away on my phone and laptop. So I spent time organising them into files and making notes to go with them, while the memories were still fresh in my mind. I knew that one day in the future, I would want to look back on them.

I remembered so many things as if they happened yesterday. Finding Nala on the mountaintop in Bosnia. Our early days together in Montenegro and Albania, our time in Santorini, and our travels through Turkey.

Then there were the people we had met. It was such a cast of characters. All thanks to Nala. The man with whom I shared an orange in the refugee camp, the family who helped me down off the mountain in Turkey, Tony and Pablo, Jason and Sirem, Nick, Iliana and Lydia; I had met so many wonderful people, some of whom I knew would be friends for life. I would never forget the debt of thanks I owe them.

Most of all, I remembered the moments I had shared with Nala. The good and the bad, the funny and the frightening.

Sifting through the photos made me appreciate her all over

again. She had taught me so much. How to enjoy the most precious moments in life. How to be myself sometimes; to do what I feel like doing and tune out everything else. But how to be of use, too. She had paved the way for me to help a lot of people, and I intended to carry on doing so.

And of course, she had taught me a lot about friendship.

A good friend isn't there all the time, but they are there when it counts. I like to think I had been there for Nala. She had certainly been there for me when the chips had been down, in Santorini and out in the middle of nowhere in Azerbaijan. I would never forget that night in the mountains in Turkey when she alerted me to the bear, or whatever it was, stalking us outside the tent. Where would I be without her?

She had been a positive influence on me in every way. I felt like I was a wiser, calmer, more mature person than the slightly wild character who left Dunbar more than a year and a half earlier. She had settled me down, made me more thoughtful, and relaxed me at times when I'd been stressed out. She was doing it again now, during our lockdown here in Hungary.

Curled up alongside me, she had accepted her isolation as a fact of life. She was simply getting on with it as she always did. So I decided to do the same, to chill out and hunker down. What else could I do? Even the most powerful people on the planet seemed powerless. There was little point in fighting our circumstances.

I had time now to think about the bigger picture and the things that really mattered. I set off from Scotland wanting to see the world in all its complexity. I had visited only eighteen countries so far, less than ten per cent of the world's nations. I had seen huge differences, but so many similarities, too. When it came down to it, we are all wrestling with the same issues, motivated by the same things. Never more so than during this pandemic.

It didn't matter now whether I was in Hungary or Hawaii, Dunbar or Durban, I was in the same boat as everyone else. It was proof, if I needed it, that we really are all in this together. I hope that when we finally overcome coronavirus and the world looks back on what it taught us, that will be the biggest lesson. It is one planet, we are one species. If we don't think about each other and work together, we're probably doomed.

As for our trip, I was used to detours and lockdowns. We'd had them before – in Montenegro and Albania, in Greece, Bulgaria and Georgia. I'd had to ride out other storms, I would ride this one out, too. In the meantime, I was able to keep distributing money to charities, some of which were going to need it more than ever.

I still believed Nala and I could find a way to get round the world together. I no longer cared which way we travelled, or how long it might take us. This far into the journey, I was crystal clear on what is and what isn't important.

At Christmas, a friend from home sent me a little book of travel quotes. I enjoyed dipping into it now and again and there were a couple of sayings I particularly liked.

The first was: *A good traveller has no fixed plans and is not intent on arriving.*

I couldn't agree more.

In the time I had been on the road, I had come to see that there was rarely much point in making too many firm plans. If I had learned anything, it was to expect the unexpected. The past few months had certainly underlined that.

The second quote was from Ernest Hemingway, who wrote: *Never go on trips with anyone you do not love.*

I understood that, too.

In Nala, I was blessed with the perfect travelling companion. I loved her not only for the joy I drew from having her alongside me on the bike each day. Or the way she allowed me to see the world through her eyes. I loved her because of what she had added to my life. Because of the meaning she'd given it – the new sense of responsibility, the purpose and direction she'd brought me. She had set me on the right path.

In which direction we'd end up heading was anyone's guess. North, south, east, west?

Fate, as always, would play its part. But then that was as it should be. It had done so ever since our paths first crossed.

Whit's fur ye'll no go past ye. We'd see where the road took us. As long as we stuck together, I knew we'd be fine.

Acknowledgements

In its own way, writing this book was almost as demanding as some of the challenges that I have had to face on my travels – but only almost! In this book, I feel that, although Nala and I are the main characters, many people have crossed our paths around the world and have played their supporting role in our story. I would also like to thank all those who helped me along the road to publication.

The idea for a book came about soon after my life had been turned upside-down by the release of the Dodo video of me finding Nala back in Bosnia. Never in my wildest dreams would I ever have thought that I could write a book, especially not one about me, and I was curious when the writer Garry Jenkins came to meet me on Santorini. Over a couple of beers watching the fire ceremony that takes place in the little village of Megalochori every summer, we agreed to collaborate. Since then, the team has expanded to include my agent Lesley Thorne at Aitken Alexander and the publishing professionals led by Rowena Webb at Hodder & Stoughton in London and Elizabeth Kulhanek at Grand Central Publishing in New York. This book would not have been possible without them and their dedicated staff – thank you! A huge thanks goes to my family and friends around the globe, who have accompanied me every step of the way – you know who you are.

Last but not least, of course, I have to thank Nala, for being the best travelling companion a human could ever wish for. She has put up with my singing, my cooking and any irritating habits I might not realise I have. I am one lucky man! Thanks, Nala, for waiting for me on that mountain that morning.

Dean Nicholson, somewhere in Europe
July 2020

Picture Acknowledgements

We would like to thank the following for their kind permission to reproduce their photographs: **Inset 1,** p. 1, top left: Neil Nicholson; middle: Gill Last; p. 3, bottom: Arbër and Kornelia from Dog Walking and Coaching, Tirana; p. 7, top right: Melina Katri. **Inset 2,** p. 5, bottom: Pablo Calvo. **Endpapers**, Nala on rug (second row, first photo): Kathrin Mormann. Picture research by Jane Smith Media.

All other pictures were taken by the author Dean Nicholson from his own collection.

For a list of the charities we have supported
with our fundraising efforts so far, visit
www.1bike1world.com/supported-charities/

Don't forget to follow us for
the latest news, stories and pictures:

www.1bike1world.com

@1bike1world

/1bike1world

@1bike1world_

An invitation from the publisher

Join us at www.hodder.co.uk, or follow us
on Twitter @hodderbooks to be a part of
our community of people who love the very
best in books and reading.

Whether you want to discover more about a book
or an author, watch trailers and interviews, have the
chance to win early limited editions, or simply browse
our expert readers' selection of the very best books,
we think you'll find what you're looking for.

And if you don't, that's the place to tell us what's missing.

We love what we do, and we'd love you to be a part of it.

www.hodder.co.uk

@hodderbooks

HodderBooks

HodderBooks